Lecture Notes
in Control and Information Sciences 390

Editors: M. Thoma, F. Allgöwer, M. Morari

T0143119

Graziano Chesi, Andrea Garulli, Alberto Tesi,
Antonio Vicino

Homogeneous Polynomial Forms for Robustness Analysis of Uncertain Systems

 Springer

Authors

Graziano Chesi, PhD

University of Hong Kong
Department of Electrical and
Electronic Engineering
Chow Yei Ching Building
Pokfulam Road
Hong Kong
P.R. China
E-mail: chesi@eee.hku.hk

Alberto Tesi, PhD

Università di Firenze
Facoltà di Ingegneria
Dipartimento di Sistemi e Informatica
Via di Santa Marta 3
50139 Firenze
Italy
E-mail: atesi@dsi.unifi.it

Andrea Garulli, PhD

Università di Siena
Dipartimento di Ingegneria
dell'Informazione
Via Roma 56
53100 Siena
Italy
E-mail: garulli@ing.unisi.it

Antonio Vicino, PhD

Università di Siena
Dipartimento di Ingegneria
dell'Informazione
Via Roma 56
53100 Siena
Italy
E-mail: vicino@ing.unisi.it

ISBN 978-1-84882-780-6   e-ISBN 978-1-84882-781-3

DOI 10.1007/978-1-84882-781-3

Lecture Notes in Control and Information Sciences   ISSN 0170-8643

Library of Congress Control Number: Applied for

©2009 Springer-Verlag Berlin Heidelberg

MATLAB® and Simulink® are registered trademarks of The MathWorks, Inc., 3 Apple Hill Drive, Natick, MA 01760-2098, U.S.A. http://www.mathworks.com

Typeset & Cover Design: Scientific Publishing Services Pvt. Ltd., Chennai, India.

Printed in acid-free paper

5 4 3 2 1 0

springer.com

To Shing Chee and Isabella (GC)

To Maria, Giovanni, and Giacomo (AG)

To Nicoletta, Andrea, and Bianca (AT)

To Maria Teresa and Ludovico (AV)

Preface

It is well known that a large number of problems relevant to the control field can be formulated as optimization problems. For long time, the classical approach has been to look for a closed form solution to the specific optimization problems at hand. The last decade has seen a noticeable shift in the meaning of "closed form" solution. The formidable increase of computational power has dramatically changed the feeling of theoreticians as well as of practitioners about what is meant by tractable and untractable problems. A main issue regards *convexity*. From a theoretical viewpoint, there has been a large amount of work in the directions of "convexifying" nonconvex problems and studying structural features of convex problems. On the other hand, extremely powerful algorithms for the solution of convex problems have been devised in the last two decades. Clearly, the fact that a wide variety of engineering problems can be formulated as convex problems has strongly motivated efforts in this direction. The control field is not an exception in this sense: many problems in robust control, identification and nonlinear control have been recognized as convex problems. Moreover, convex relaxations of nonconvex problems have been intensively investigated, as they provide an effective tool for bounding the optimal solution of the original problem.

As far as robust control is concerned, it is known since long time that several classes of problems can be reduced to testing positivity of suitable polynomials. Remarkable examples are: the construction of Lyapunov functions and the evaluation of the stability margin for systems affected by structured uncertainty; the estimation of the domain of attraction of nonlinear systems; the synthesis of fixed-order $\mathcal{H}_\infty$ optimal controllers; the robust disturbance rejection for nonlinear systems, and many others. In recent years, it has been recognized that positivity of polynomials can be tackled effectively in the framework of linear matrix inequality (LMI) problems, which are a special class of convex optimizations problems enjoying a number of appealing properties, like solvability in polynomial time. A fundamental relaxation consists of replacing the condition for a polynomial to be positive with the condition that it is a sum of squares of polynomials (SOS). The interest for this relaxation is motivated by the facts that testing whether a polynomial is SOS boils down to an LMI problem, and that the conservatism of this relaxation can be reduced by suitably increasing the degree of the polynomial.

This book describes some techniques developed recently by the authors and a number of researchers in the field, in order to address stability and performance analysis of uncertain systems affected by structured parametric uncertainties. Convex relaxations for different robustness problems are constructed by employing *homogeneous forms* (hereafter, simply *forms*), i.e. polynomials whose terms have the same degree. Such forms are used to parametrize various classes of Lyapunov functions. The proposed solutions are characterized in terms of LMI problems and show a number of nice theoretical and practical features, which are illustrated throughout the book.

Organization of the Book

Chapter 1 introduces the square matricial representation (SMR), which is a powerful tool for the representation of forms as it allows one to establish whether a form is SOS via an LMI feasibility test. This tool is then extended to address the representation of matrix forms and the characterization of SOS matrix forms. It is shown that, for a given form, an SOS index can be computed by solving an eigenvalue problem (EVP), which is the minimization of a linear function subject to LMI constraints, also known as semidefinite program. Moreover, it is shown how the positivity of a polynomial on an ellipsoid and the positivity of a matrix polynomial on the simplex can be cast in terms of positivity of suitable forms. The chapter also provides conditions under which vectors related to a given homogeneous polynomial structure lie in assigned linear spaces. This is useful in order to study the conservatism of SOS-based relaxations.

Chapter 2 investigates the relationship between convex relaxations for positivity of forms and Hilbert's 17th problem, which concerns the existence of positive forms which are not SOS forms (PNS). The concepts of maximal SMR matrices and SMR-tight forms are introduced, which allow one to derive *a posteriori* tightness conditions for LMI optimizations arising in SOS relaxations. Also, this chapter provides results on Hilbert's 17th problem based on the SMR, showing that each PNS form is the vertex of a cone of PNS forms, and providing a parametrization of the set of PNS forms.

Chapter 3 addresses robust stability and robust performance analysis of time-varying polytopic systems, i.e. uncertain systems affected by linear dependent time-varying uncertainties constrained in a polytope. It is shown how robustness properties can be investigated by using homogeneous polynomial Lyapunov functions (HPLFs), a non-conservative class of Lyapunov functions whose construction can be tackled by solving LMI problems such as LMI feasibility tests or generalized eigenvalue problems (GEVPs), being the latter a class of quasi-convex optimizations with LMI constraints. Moreover, *a priori* conditions for tightness of the considered relaxations are provided. The extension to the case of uncertain systems with rational dependence on the uncertain parameters is derived through linear fractional representations (LFRs).

Chapter 4 investigates robustness analysis of time-invariant polytopic systems by adopting homogeneous parameter-dependent quadratic Lyapunov functions (HPD-QLFs), again a non-conservative class of Lyapunov functions. The chapter provides *a posteriori* tests for establishing non-conservatism of the bounds obtained for robust stability margin or robust performance. Alternative conditions for assessing robust stability and instability of time-invariant polytopic systems are provided through LMI optimizations resulting from classical stability criteria. Moreover, it is shown how such results can be extended to the case of uncertain systems with rational dependence on the uncertain parameter, and to the case of discrete-time systems.

Chapter 5 considers the case of polytopic systems with time-varying uncertainties and finite bounds on their variation rate, and illustrates how robustness analysis can be addressed by using homogeneous parameter-dependent homogeneous Lyapunov functions (HPD-HLFs). This class of functions include all possible forms in the state variables and uncertain parameters, and therefore recovers the classes of HPLFs and HPD-QLFs as special cases. The chapter shows how the construction of HPD-QLFs can be formulated in terms of LMI problems, and highlights the role played by the degrees of the Lyapunov function in the state variables and uncertain parameters.

Lastly, Chapter 6 treats quadratic distance problems (QDPs), i.e. the computation of the minimum weighted euclidean distance from a point to a surface defined by a polynomial equation. This special class of nonconvex optimization problems finds numerous applications in control systems. It is shown that a lower bound to the solution of a QDP can be obtained through a sequence of LMI feasibility tests. *a priori* and *a posteriori* necessary and sufficient conditions for tightness of the lower bound are derived. The proposed technique is applied to the computation of the parametric stability margin of time-invariant polytopic systems.

Acknowledgements

We thank various colleagues for the fruitful discussions in last years that contributed to the composition of this book, in particular F. Blanchini, P.-A. Bliman, S. Boyd, P. Colaneri, Y. Ebihara, Y. Fujisaki, M. Fujita, R. Genesio, D. Henrion, Y. S. Hung, J.-B. Lasserre, J. Lofberg, M. Khammash, L. Ljung, A. Masi, S. Miani, Y. Oishi, C. Papageorgiou, S. Prajna, M. Sato, C. W. Scherer and A. Seuret.

This book was typeset by the authors using LATEX. All computations were done using MATLAB® and SeDuMi.

As much as we wish the book to be free of errors, we know this will not be the case. Therefore, we will greatly appreciate to receive reports of errors, which can be sent to the following address: `chesi@eee.hku.hk`. An up-to-date errata list will be available at the following homepage: `http://www.eee.hku.hk/~chesi`.

Hong Kong   *Graziano Chesi*
Siena   *Andrea Garulli*
Firenze   *Alberto Tesi*
Siena   *Antonio Vicino*
May 2009

Contents

Notation

Basic Sets

$\mathbb{N}$ space of natural numbers (including zero)
$\mathbb{R}$ space of real numbers
$\mathbb{C}$ space of complex numbers
$[a,b]$ $\{x \in \mathbb{R} : a \leq x \leq b\}$, with $a, b \in \mathbb{R}$
0_n $n \times 1$ null vector
$0_{m \times n}$ $m \times n$ null matrix
I_n $n \times n$ identity matrix
$\mathbb{R}_0^n$ $\mathbb{R}^n \setminus \{0_n\}$
$\mathbb{S}^n$ set of $n \times n$ real symmetric matrices

Basic Functions and Operators

$\star$ lower triangular entries in symmetric matrices
$\mathrm{re}(\lambda)$ real part of $\lambda \in \mathbb{C}$
$\mathrm{im}(\lambda)$ imaginary part of $\lambda \in \mathbb{C}$
$|\lambda|$ $\sqrt{\mathrm{re}(\lambda)^2 + \mathrm{im}(\lambda)^2}$, with $\lambda \in \mathbb{C}$
x^y $x_1^{y_1} x_2^{y_2} \cdots x_n^{y_n}$, with $x \in \mathbb{R}^n$, $y \in \mathbb{R}^n$
$\mathrm{sq}(x)$ $(x_1^2, \ldots, x_n^2)'$, with $x \in \mathbb{R}^n$
$\mathrm{sqr}(x)$ $(\sqrt{x_1}, \ldots, \sqrt{x_n})'$, with $x \in \mathbb{R}^n$
$\|x\|$ 2-norm of $x \in \mathbb{R}^n$, i.e. $\|x\| = \sqrt{x'x}$
$\|x\|_\infty$ ∞-norm of $x \in \mathbb{R}^n$, i.e. $\|x\|_\infty = \max_{i=1,\ldots,n} |x_i|$
$x > 0$ $x_i > 0 \ \forall i = 1, \ldots, n$, with $x \in \mathbb{R}^n$
$x \geq 0$ $x_i \geq 0 \ \forall i = 1, \ldots, n$, with $x \in \mathbb{R}^n$
$\det(X)$ determinant of $X \in \mathbb{R}^{n \times n}$
$\mathrm{rank}(X)$ rank of $X \in \mathbb{R}^{m \times n}$
X' transpose of $X \in \mathbb{R}^{m \times n}$
$\mathrm{he}(X)$ $X + X'$, with $X \in \mathbb{R}^{n \times n}$
$\mathrm{diag}(x)$ $n \times n$ diagonal matrix where the entry (i, i) is x_i, with $x \in \mathbb{R}^n$

$\text{co}\{X_1,\ldots,X_p\}$ convex hull of matrices $X_1,\ldots,X_p \in \mathbb{R}^{m\times n}$, i.e.

$$\text{co}\{X_1,\ldots,X_p\} = \left\{ X = \sum_{i=1}^{p} y_i X_i, \ \ y_i \in [0,1], \ \sum_{i=1}^{p} y_i = 1 \right\}$$

$\text{img}(X)$ image of $X \in \mathbb{R}^{m\times n}$, i.e.

$$\text{img}(X) = \{y \in \mathbb{R}^m : y = Xz, \ z \in \mathbb{R}^n\}$$

$\text{ker}(X)$ right null space of $X \in \mathbb{R}^{m\times n}$, i.e.

$$\text{ker}(X) = \{z \in \mathbb{R}^n : Xz = 0_m\}$$

$\text{spc}(X)$ set of eigenvalues of $X \in \mathbb{R}^{n\times n}$, i.e.

$$\text{spc}(X) = \{\lambda \in \mathbb{C} : \ \det(\lambda I_n - X) = 0\}$$

$\lambda_{max}(X)$ maximum real eigenvalue of $X \in \mathbb{R}^{n\times n}$
$\lambda_{min}(X)$ minimum real eigenvalue of $X \in \mathbb{R}^{n\times n}$
$X \otimes Y$ Kronecker product of matrices X and Y, i.e.

$$X \otimes Y = \begin{pmatrix} x_{1,1}Y & x_{1,2}Y & \cdots \\ x_{2,1}Y & x_{2,2}Y & \cdots \\ \vdots & \vdots & \ddots \end{pmatrix}, \ \ X = \begin{pmatrix} x_{1,1} & x_{1,2} & \cdots \\ x_{2,1} & x_{2,2} & \cdots \\ \vdots & \vdots & \ddots \end{pmatrix}$$

$X^{[i]}$ i-th Kronecker power, i.e.

$$X^{[i]} = \begin{cases} X \otimes X^{[i-1]} & \text{if } i > 1 \\ 1 & \text{if } i = 0 \end{cases}$$

$X > 0$ symmetric positive definite matrix $X \in \mathbb{S}^n$, i.e.

$$X > 0 \iff y'Xy > 0 \ \forall y \in \mathbb{R}_0^n$$

$X \geq 0$ symmetric positive semidefinite matrix $X \in \mathbb{S}^n$, i.e.

$$X \geq 0 \iff y'Xy \geq 0 \ \forall y \in \mathbb{R}^n$$

Abbreviations

EVP	Eigenvalue problem
GEVP	Generalized eigenvalue problem
HPD-HLF	Homogeneous parameter-dependent homogeneous Lyapunov function
HPD-QLF	Homogeneous parameter-dependent quadratic Lyapunov function
HPLF	Homogeneous polynomial Lyapunov function
LFR	Linear fractional representation
LMI	Linear matrix inequality
PNS	Positive non SOS
SMR	Square matricial representation
SOS	Sum of squares of polynomials

Chapter 1
Positive Forms

This chapter addresses the problem of studying positivity of a *form*, i.e. a polynomial whose terms have all the same degree. This is a key issue which has many implications in systems and control theory. A basic tool for the representation of forms, which is known in the literature as Gram matrix or SMR, is introduced. The main idea is to represent a form of a generic degree through a quadratic form, by introducing suitable base vectors and corresponding coefficient matrices, which are called power vectors and SMR matrices, respectively. It is shown that a positive semidefinite SMR matrix exists if and only if the form is an SOS form. This allows one to establish whether a form is SOS via an LMI feasibility test, which is a convex optimization problem. Hence, sufficient conditions for positivity of forms can be formulated in terms of LMIs. Then, the SMR framework is extended to address the case of matrix forms. Another contribution of this chapter is to show how some problems involving positivity of polynomials over special sets, such as ellipsoids or the simplex, can be cast in terms of unconstrained positivity of a form. Finally, it is shown how the power vectors belonging to assigned linear spaces can be determined via linear algebra operations. This is instrumental to the extraction of solutions in the robustness problems addressed in subsequent chapters.

1.1 Forms and Polynomials

Let us define the set

$$\mathscr{D}_{n,d} = \left\{ q \in \mathbb{N}^n : \sum_{i=1}^{n} q_i = d \right\}. \tag{1.1}$$

Definition 1.1 (Form). The function $h : \mathbb{R}^n \to \mathbb{R}$ is a *form* of degree d in n scalar variables if

$$h(x) = \sum_{q \in \mathscr{D}_{n,d}} a_q x^q \tag{1.2}$$

where $x \in \mathbb{R}^n$ and $a_q \in \mathbb{R}$ is the coefficient of the monomial x^q.

G. Chesi et al.: Homogeneous Polynomial Forms, LNCIS 390, pp. 1–37.
springerlink.com   © Springer-Verlag Berlin Heidelberg 2009

Hence, forms are weighted sums of monomials of the same degree. We define the set of forms of degree d in n scalar variables as

$$\Xi_{n,d} = \{h : \mathbb{R}^n \to \mathbb{R} \ : \ (1.2) \text{ holds}\}. \tag{1.3}$$

Let us now introduce the family of parametrized forms. Let $h(x; p)$ be a form of degree d in $x \in \mathbb{R}^n$ for any fixed parameter vector p in some space. Then, the notation

$$h(\cdot; p) \in \Xi_{n,d}$$

will denote that $h(\cdot; p)$ is a family of forms parametrized by p.

Generic polynomials can be represented by using forms, according to the following definition.

Definition 1.2 (Polynomial). The function $f : \mathbb{R}^n \to \mathbb{R}$ is a *polynomial* of degree less than or equal to d, in n scalar variables, if

$$f(x) = \sum_{i=0}^{d} h_i(x) \tag{1.4}$$

where $x \in \mathbb{R}^n$ and $h_i \in \Xi_{n,i}$, $i = 1, \ldots, d$.

This means that polynomials are sums of forms, i.e. forms are special cases of polynomials. At the same time, any polynomial $f(x)$ can be obtained by restricting a suitable form $\hat{h}(y)$ to a subspace, e.g. according to

$$f(x) = \hat{h}(y)|_{y_{n+1}=1},$$

where $y = (x', 1)'$ and

$$\hat{h}(y) = \sum_{i=0}^{d} h_i(x) y_{n+1}^{d-i}.$$

1.2 Representation via Power Vectors

Forms can be represented by vectors which contain their coefficients with respect to an appropriate base. First of all, let us observe that the number of coefficients of any form in $\Xi_{n,d}$ is given by the cardinality of $\mathscr{Q}_{n,d}$, which is equal to

$$\sigma(n,d) = \frac{(n+d-1)!}{(n-1)!d!}. \tag{1.5}$$

Definition 1.3 (Power Vector). Let $x^{\{d\}}$ be any vector in $\mathbb{R}^{\sigma(n,d)}$ such that, for all $h \in \Xi_{n,d}$, there exists $g \in \mathbb{R}^{\sigma(n,d)}$ satisfying

$$h(x) = g'x^{\{d\}}. \tag{1.6}$$

Then, $x^{\{d\}}$ is called a *power vector* for $\Xi_{n,d}$.

Therefore, $x^{\{d\}}$ is a vector of forms of degree d in x whose entries are a finite generating set for $\Xi_{n,d}$, i.e. every form $h \in \Xi_{n,d}$ can be represented as a linear combination of the elements of $x^{\{d\}}$, according to (1.6).

Special choices for $x^{\{d\}}$ are those where each entry is a monomial. This means that

$$\left(x^{\{d\}}\right)_i = x^{\varphi(i)} \tag{1.7}$$

where $\left(x^{\{d\}}\right)_i$ is the i-th entry of $x^{\{d\}}$, and φ is any bijective function such that

$$\varphi : \{i \in \mathbb{N} : 1 \leq i \leq \sigma(n,d)\} \to \mathscr{D}_{n,d}.$$

Among these choices, a typical one is

$$\begin{pmatrix} x_1 \\ x_2 \\ \vdots \\ x_n \end{pmatrix}^{\{k\}} = \begin{cases} \begin{pmatrix} x_1 (x_1, x_2, x_3, \ldots, x_n)'^{\{k-1\}} \\ x_2 (x_2, x_3, \ldots, x_n)'^{\{k-1\}} \\ \vdots \\ x_n (x_n)'^{\{k-1\}} \end{pmatrix} & \text{if } k > 0 \\ 1 & \text{otherwise} \end{cases} \tag{1.8}$$

which yields the lexicographical order of the monomials in $x^{\{d\}}$. Other choices of interest for $x^{\{d\}}$ are those satisfying the property

$$x^{\{d\}'} x^{\{d\}} = \|x\|^{2d}. \tag{1.9}$$

For instance, this can be obtained by weighting the entries of $x^{\{d\}}$ in (1.7) as follows:

$$\left(x^{\{d\}}\right)_i = \sqrt{\frac{d!}{(\varphi(i))_1! (\varphi(i))_2! \cdots (\varphi(i))_n!}} \, x^{\varphi(i)}. \tag{1.10}$$

Example 1.1. Let us consider the form

$$h(x) = x_1^3 + 2x_1^2 x_2 - 4x_2^3.$$

One has that $h \in \Xi_{2,3}$, i.e. $n = 2$ and $d = 3$. Then, $h(x)$ can be written as in (1.6) with

$$g = \begin{pmatrix} 1 \\ 2 \\ 0 \\ -4 \end{pmatrix}, \quad x^{\{d\}} = \begin{pmatrix} x_1^3 \\ x_1^2 x_2 \\ x_1 x_2^2 \\ x_2^3 \end{pmatrix}$$

where the chosen $x^{\{3\}}$ is a power vector for $\Xi_{2,3}$. Observe that the dimension of $x^{\{d\}}$ is given by (1.5) and it is equal to $\sigma(2,3) = 4$. Observe also that this choice for $x^{\{3\}}$ does not satisfy (1.9), indeed

$$
\begin{aligned}
x^{\{d\}'} x^{\{d\}} &= x_1^6 + x_1^4 x_2^2 + x_1^2 x_2^4 + x_2^6 \\
&\neq x_1^6 + 3x_1^4 x_2^2 + 3x_1^2 x_2^4 + x_2^6 \\
&= \|x\|^{2d}.
\end{aligned}
$$

Alternatively, one can choose $x^{\{d\}}$ satisfying (1.9), as suggested in (1.10), thus obtaining e.g.

$$
g = \begin{pmatrix} 1 \\ 2 \cdot 3^{-\frac{1}{2}} \\ 0 \\ -4 \end{pmatrix}, \quad x^{\{d\}} = \begin{pmatrix} x_1^3 \\ \sqrt{3} x_1^2 x_2 \\ \sqrt{3} x_1 x_2^2 \\ x_2^3 \end{pmatrix}.
$$

1.3 Representation via SMR

Forms of even degree can be also represented via matrices. Indeed, let us observe that $x^{\{m\}'} H x^{\{m\}}$ is a form of degree $2m$ in x for all $H \in \mathbb{S}^{\sigma(n,m)}$. This suggests that a form $h \in \Xi_{n,2m}$ can be represented via a suitable matrix $H \in \mathbb{S}^{\sigma(n,m)}$.

Definition 1.4 (SMR). Let $h \in \Xi_{n,2m}$ and $H \in \mathbb{S}^{\sigma(n,m)}$ be such that

$$
h(x) = x^{\{m\}'} H x^{\{m\}}. \tag{1.11}
$$

Then, (1.11) is called a *SMR* of $h(x)$ with respect to $x^{\{m\}}$. Moreover, H is called a *SMR matrix* of $h(x)$ with respect to $x^{\{m\}}$.

The existence and parametrization of the matrices H fulfilling (1.11) is considered hereafter.

1.3.1 Equivalent SMR Matrices

The following result clarifies that any form in $\Xi_{n,2m}$ can be expressed as in (1.11).

Theorem 1.1. *Let $h \in \Xi_{n,2m}$. Then, for any power vector $x^{\{m\}}$ there exists $H \in \mathbb{S}^{\sigma(n,m)}$ such that (1.11) holds.*

Proof. Let $a(x) \in \mathbb{R}^{\sigma(n,m)}$ be any power vector for $\Xi_{n,m}$ such that its entries are monomials of degree m in x. One can always write

$$h(x) = a(x)'Ba(x) \tag{1.12}$$

where $B \in \mathbb{S}^{\sigma(n,m)}$. Indeed, let us observe that $(a(x))_i B_{i,j} (a(x))_j$ is a monomial of degree $2m$ in x for each pair (i,j). Moreover, for each monomial of degree $2m$ in x there exists at least one pair (i,j) such that $(a(x))_i B_{i,j} (a(x))_j$ is equal to this monomial up to a scalar factor. Therefore, for any $h \in \Xi_{n,2m}$ the matrix B satisfying (1.12) can be simply built as follows. First, for each monomial of $h(x)$, set one of the entries of B corresponding to this monomial equal to its coefficient in $h(x)$ and set to zero all the other entries of B corresponding to the same monomial. Second, replace B by $\mathrm{he}(B)/2$.

Now, the matrix H satisfying (1.11) can be obtained from B as

$$H = (C^{-1})'BC^{-1}$$

where $C \in \mathbb{R}^{\sigma(n,m) \times \sigma(n,m)}$ is a nonsingular matrix such that

$$x^{\{m\}} = Ca(x). \tag{1.13}$$

This matrix C clearly exists since $x^{\{m\}}$ and $a(x)$ are power vectors for the forms of degree m in x. $\qquad \square$

Let us consider the problem of characterizing the set of matrices H satisfying (1.11). We define such a set as

$$\mathscr{H}(h) = \left\{ H \in \mathbb{S}^{\sigma(n,m)} : (1.11) \text{ holds} \right\}. \tag{1.14}$$

The following result states an important property of $\mathscr{H}(h)$.

Lemma 1.1. *Let* $h \in \Xi_{n,2m}$. *Then,* $\mathscr{H}(h)$ *is an affine space.*

Proof. Let H_1 and H_2 be any matrices in $\mathscr{H}(h)$, and let us define $H_0 = a_1 H_1 + a_2 H_2$ for some $a_1, a_2 \in \mathbb{R}$. We have that

$$\begin{aligned} x^{\{m\}'} H_0 x^{\{m\}} &= x^{\{m\}'} (a_1 H_1 + a_2 H_2) x^{\{m\}} \\ &= a_1 x^{\{m\}'} H_1 x^{\{m\}} + a_2 x^{\{m\}'} H_2 x^{\{m\}} \\ &= (a_1 + a_2) h(x). \end{aligned}$$

This implies that

$$H_0 \in \mathscr{H}(h) \quad \forall a_1, a_2 \in \mathbb{R} : a_1 + a_2 = 1$$

which means that $\mathscr{H}(h)$ is an affine space. $\qquad \square$

Since $\mathscr{H}(h)$ is an affine space, its elements can be expressed as the sum between any element of $\mathscr{H}(h)$ itself and an element free to vary in a suitable linear space. Indeed, one has that

$$\mathscr{H}(h) = \left\{ H + L, \quad H \in \mathbb{S}^{\sigma(n,m)} \text{ such that (1.11) holds}, L \in \mathscr{L}_{n,m} \right\}$$

where $\mathscr{L}_{n,m}$ is the linear space

$$\mathscr{L}_{n,m} = \left\{ L \in \mathbb{S}^{\sigma(n,m)} : \ x^{\{m\}'} L x^{\{m\}} = 0 \ \forall x \in \mathbb{R}^n \right\}. \tag{1.15}$$

The following result characterizes the dimension of $\mathscr{L}_{n,m}$.

Theorem 1.2. *The dimension of $\mathscr{L}_{n,m}$ is given by*

$$\omega(n,m) = \frac{1}{2}\sigma(n,m)\left(1 + \sigma(n,m)\right) - \sigma(n,2m). \tag{1.16}$$

Proof. Let us define

$$a = \frac{1}{2}\sigma(n,m)\left(1 + \sigma(n,m)\right)$$

and let $b \in \mathbb{R}^a$ be a vector containing the free entries of a matrix $L \in \mathbb{S}^{\sigma(n,m)}$. Let $C : \mathbb{R}^a \to \mathbb{S}^{\sigma(n,m)}$ be the linear map from b to the corresponding L, i.e.

$$C(b) = L.$$

Let us write

$$\begin{aligned} x^{\{m\}'} L x^{\{m\}} &= x^{\{m\}'} C(b) x^{\{m\}} \\ &= (Db)' x^{\{2m\}} \end{aligned}$$

where $D \in \mathbb{R}^{\sigma(n,2m) \times a}$ is a suitable matrix. It follows that

$$\mathscr{L}_{n,m} = \{C(b) : \ b \in \ker(D)\}.$$

This implies that

$$\begin{aligned} \dim\left(\mathscr{L}_{n,m}\right) &= \dim\left(\{C(b) : \ b \in \ker(D)\}\right) \\ &= \dim\left(\ker(D)\right) \\ &= a - \mathrm{rank}(D). \end{aligned}$$

Now, let us prove that

$$\mathrm{rank}(D) = \sigma(n,2m).$$

In fact, let us suppose by contradiction that $\mathrm{rank}(D) \neq \sigma(n,2m)$. Since $\sigma(n,2m) \leq a$, one has that $\mathrm{rank}(D) < \sigma(n,2m)$. But this in turn implies that

$$\exists h \in \Xi_{n,2m} : \ (Db)' x^{\{2m\}} \neq h(x) \ \ \forall b \in \mathbb{R}^a$$

or, in other words

$$\exists h \in \Xi_{n,2m} : \ x^{\{m\}'} L x^{\{m\}} \neq h(x) \ \ \forall L \in \mathbb{S}^{\sigma(n,m)}$$

which contradicts Theorem 1.1. Therefore, $\mathrm{rank}(D) = \sigma(n, 2m)$, and hence the dimension of $\mathscr{L}_{n,m}$ is given by $\omega(n,m)$ in (1.16). □

It can be verified that $\mathscr{L}_{n,m}$ contains only one matrix in the cases $n = 1$ for all m, and $m = 1$ for all n. Indeed, one has

$$\omega(1,m) = 0$$
$$\omega(n,1) = 0.$$

Table 1.1 shows the quantities $\sigma(n,m)$ and $\omega(n,m)$ given by (1.5) and (1.16), respectively, for some values of n, m.

Table 1.1 Quantities $\sigma(n,m)$ and $\omega(n,m)$ for some values of n, m: (a) $\sigma(n,m)$; (b) $\omega(n,m)$

| | (a) | | | | | | (b) | | | | |
	$m = 1$	$m = 2$	$m = 3$	$m = 4$	$m = 5$		$m = 1$	$m = 2$	$m = 3$	$m = 4$	$m = 5$
$n = 1$	1	1	1	1	1	$n = 1$	0	0	0	0	0
$n = 2$	2	3	4	5	6	$n = 2$	0	1	3	6	10
$n = 3$	3	6	10	15	21	$n = 3$	0	6	27	75	165
$n = 4$	4	10	20	35	56	$n = 4$	0	20	126	465	1310
$n = 5$	5	15	35	70	126	$n = 5$	0	50	420	1990	7000

1.3.2 Complete SMR

Let $L(\alpha)$ be a linear parametrization of $\mathscr{L}_{n,m}$ where $\alpha \in \mathbb{R}^{\omega(n,m)}$ is a free vector. The set $\mathscr{H}(h)$ in (1.14) can be rewritten as

$$\mathscr{H}(h) = \left\{ H + L(\alpha) : H \in \mathbb{S}^{\sigma(n,m)} \text{ satisfies (1.11)}, \ \alpha \in \mathbb{R}^{\omega(n,m)} \right\}.$$

This allows us to derive the following parametrization for forms of even degree.

Definition 1.5 (Complete SMR). Consider any $h \in \Xi_{n,2m}$. Let $H \in \mathbb{S}^{\sigma(n,m)}$ be such that (1.11) holds, and $L(\alpha)$ be a linear parametrization of $\mathscr{L}_{n,m}$ with $\alpha \in \mathbb{R}^{\omega(n,m)}$. Then, $h(x)$ can be written as

$$h(x) = x^{\{m\}'} (H + L(\alpha)) x^{\{m\}}. \tag{1.17}$$

The expression in (1.17) is called a *complete SMR* of $h(x)$ with respect to $x^{\{m\}}$. Moreover, $H + L(\alpha)$ is called a *complete SMR matrix* of $h(x)$ with respect to $x^{\{m\}}$.

Algorithms for the construction of the matrices H and $L(\alpha)$ in (1.17) are reported in Appendix B.

It is worthwhile to observe that the complete SMR matrix $H + L(\alpha)$ is not unique since the choice for H and $L(\alpha)$ is not unique. Also, $H + L(\alpha)$ depends on the selected power vector $x^{\{m\}}$. The following result characterizes different complete SMR matrices of $h(x)$.

Theorem 1.3. *Consider the complete SMR of $h(x)$ in (1.17). Then,*

$$h(x) = \overline{x^{\{m\}}}' \, (\bar{H} + \bar{L}(\bar{\alpha})) \, \overline{x^{\{m\}}} \tag{1.18}$$

with

$$\begin{aligned}
\overline{x^{\{m\}}} &= A^{-1} x^{\{m\}} \\
\bar{H} &= A' \, (H - L(B\beta)) \, A \\
\bar{L}(\bar{\alpha}) &= A' L(B\bar{\alpha}) A \\
\bar{\alpha} &= B^{-1}\alpha + \beta
\end{aligned} \tag{1.19}$$

is also a complete SMR of $h(x)$ for all $A \in \mathbb{R}^{\sigma(n,m) \times \sigma(n,m)}$, $B \in \mathbb{R}^{\omega(n,m) \times \omega(n,m)}$ and $\beta \in \mathbb{R}^{\omega(n,m)}$, with A and B nonsingular.

Proof. By substituting in (1.18) the expressions of $\overline{x^{\{m\}}}, \bar{H}, \bar{L}(\bar{\alpha})$ given in (1.19), one obtains

$$\begin{aligned}
h(x) &= \overline{x^{\{m\}}}' \, (\bar{H} + \bar{L}(\bar{\alpha})) \, \overline{x^{\{m\}}} \\
&= x^{\{m\}'} A^{-1'} (A'(H - L(B\beta))A + A'L(\alpha + B\beta)A) A^{-1} x^{\{m\}} \\
&= x^{\{m\}'} (H + L(\alpha)) x^{\{m\}}
\end{aligned}$$

and hence the theorem holds. □

Example 1.2. Consider the form

$$h(x) = x_1^4 + 2x_1^3 x_2 + 2x_2^4. \tag{1.20}$$

One has that $h \in \Xi_{2,4}$, i.e. $n = 2$ and $m = 2$. Then, $h(x)$ can be written as in (1.17) with

$$x^{\{m\}} = \begin{pmatrix} x_1^2 \\ x_1 x_2 \\ x_2^2 \end{pmatrix}, \quad H = \begin{pmatrix} 1 & 1 & 0 \\ \star & 0 & 0 \\ \star & \star & 2 \end{pmatrix}, \quad L(\alpha) = \begin{pmatrix} 0 & 0 & -\alpha_1 \\ \star & 2\alpha_1 & 0 \\ \star & \star & 0 \end{pmatrix}. \tag{1.21}$$

In fact, the dimension of α is given by $\omega(2,2) = 1$, according to (1.16).

Example 1.3. Consider the form

$$h(x) = x_1^4 + 2x_2^4 + 3x_3^4. \tag{1.22}$$

One has $n = 3$, $m = 2$, and (1.17) holds with

$$
x^{\{m\}} = \begin{pmatrix} x_1^2 \\ x_1 x_2 \\ x_1 x_3 \\ x_2^2 \\ x_2 x_3 \\ x_3^2 \end{pmatrix}, \quad
H = \begin{pmatrix}
1 & 0 & 0 & 0 & 0 & 0 \\
\star & 0 & 0 & 0 & 0 & 0 \\
\star & \star & 0 & 0 & 0 & 0 \\
\star & \star & \star & 2 & 0 & 0 \\
\star & \star & \star & \star & 0 & 0 \\
\star & \star & \star & \star & \star & 3
\end{pmatrix}
$$

$$
L(\alpha) = \begin{pmatrix}
0 & 0 & 0 & -\alpha_1 & -\alpha_2 & -\alpha_3 \\
\star & 2\alpha_1 & \alpha_2 & 0 & -\alpha_4 & -\alpha_5 \\
\star & \star & 2\alpha_3 & \alpha_4 & \alpha_5 & 0 \\
\star & \star & \star & 0 & 0 & -\alpha_6 \\
\star & \star & \star & \star & 2\alpha_6 & 0 \\
\star & \star & \star & \star & \star & 0
\end{pmatrix}.
$$

(1.23)

In the sequel, the power vector $x^{\{m\}}$ used to define the SMR will be always given by (1.8), unless otherwise specified.

1.4 SOS Forms

The following definition introduces a key class of forms.

Definition 1.6 (SOS Form). The form $h \in \Xi_{n,2m}$ is an *SOS form* if there exist $h_1, \ldots, h_j \in \Xi_{n,m}$ for some integer $j \geq 1$ such that

$$
h(x) = \sum_{i=1}^{j} h_i(x)^2.
\tag{1.24}
$$

We denote the set of SOS forms of degree $2m$ in n scalar variables with the notation

$$
\Sigma_{n,2m} = \{ h \in \Xi_{n,2m} : h(x) \text{ is SOS} \}.
\tag{1.25}
$$

As it will be explained in the sequel, this set can be characterized by using the SMR.

1.4.1 SOS Tests Based on the SMR

The SMR allows one to establish whether a form is SOS through a convex optimization. This is clarified in the following result.

Theorem 1.4. *Let $H + L(\alpha)$ be a complete SMR matrix of $h \in \Xi_{n,2m}$. Then, $h(x)$ is SOS if and only if*

$$\exists \alpha : H + L(\alpha) \geq 0. \tag{1.26}$$

Proof. (Necessity) Since $h(x)$ is SOS, there exist $h_1(x), \ldots, h_j(x)$ such that (1.24) holds. Let us write each $h_i(x)$ as

$$h_i(x) = g_i' x^{\{m\}}$$

where $g_1, \ldots, g_j \in \mathbb{R}^{\sigma(n,m)}$ are suitable vectors of coefficients. It follows that:

$$h(x) = \sum_{i=1}^{j} h_i(x)^2 = \sum_{i=1}^{j} x^{\{m\}'} g_i g_i' x^{\{m\}} = x^{\{m\}'} \bar{H} x^{\{m\}}$$

where

$$\bar{H} = \sum_{i=1}^{j} g_i g_i'.$$

Hence, $\bar{H}$ is an SMR matrix of $g(x)$. Moreover, it clearly holds that

$$\bar{H} \geq 0.$$

Since $H + L(\alpha)$ parametrizes all the SMR matrices of $h(x)$, there exists $\bar{\alpha}$ such that

$$H + L(\bar{\alpha}) = \bar{H}.$$

and hence (1.26) holds.

(Sufficiency) Let $\bar{\alpha}$ be a value of α satisfying (1.26). Since $H + L(\bar{\alpha}) \geq 0$, there exists a Cholesky factor $G \in \mathbb{R}^{\sigma(n,m) \times \sigma(n,m)}$ such that

$$H + L(\bar{\alpha}) = G'G.$$

Let g_i' be the i-th row of G. It follows:

$$\begin{aligned}
h(x) &= x^{\{m\}'} (H + L(\bar{\alpha})) x^{\{m\}} \\
&= x^{\{m\}'} G'G x^{\{m\}} \\
&= \begin{pmatrix} g_1' x^{\{m\}} \\ g_2' x^{\{m\}} \\ \vdots \\ g_{\sigma(n,m)}' x^{\{m\}} \end{pmatrix}' \begin{pmatrix} g_1' x^{\{m\}} \\ g_2' x^{\{m\}} \\ \vdots \\ g_{\sigma(n,m)}' x^{\{m\}} \end{pmatrix} \\
&= \sum_{i=1}^{\sigma(n,m)} h_i(x)^2
\end{aligned} \tag{1.27}$$

where $h_i(x) = g_i' x^{\{m\}}$ is a form of degree m for all $i = 1, \ldots, \sigma(n,m)$. Therefore, $h(x)$ is SOS. $\qquad\square$

From Theorem 1.4 it follows that to establish whether $h(x)$ is SOS is equivalent to establish whether the inequality (1.26) admits a feasible solution for α. Since $H + L(\alpha)$ is an affine linear matrix function of the variable α, we have that (1.26) is an LMI (see Appendix A.1 for details). Therefore, to establish whether $h(x)$ is SOS amounts to solving an LMI feasibility test, which belongs to the class of convex optimization problems.

The following corollary is a direct consequence of Theorem 1.4 and provides an upper bound to the quantity j in (1.24).

Corollary 1.1. *Let $h \in \Sigma_{n,2m}$. Then, (1.24) holds with*

$$j \leq \sigma(n,m). \tag{1.28}$$

Proof. Since $h(x)$ is SOS, from Theorem 1.4 there exists $\bar{\alpha}$ such that $H + L(\bar{\alpha}) \geq 0$. In particular, from (1.27) it turns out that $h(x) = \sum_{i=1}^{\sigma(n,m)} h_i(x)^2$. Since $h_i(x)$ could be zero for some i, it follows that (1.28) holds. $\qquad\square$

1.4.2 SOS Index

In order to further characterize SOS forms, we introduce the following definition.

Definition 1.7 (SOS Index). Let $h \in \Xi_{n,2m}$ and define

$$\begin{aligned} \lambda(h) = \max_{t,\alpha} \ & t \\ \text{s.t. } \ & H + L(\alpha) - t I_{\sigma(n,m)} \geq 0 \end{aligned} \tag{1.29}$$

where $H + L(\alpha)$ is a complete SMR matrix of $h(x)$. Then, $\lambda(h)$ is called SOS index of $h(x)$.

The optimization problem (1.29) is known in the literature as EVP and turns out to be convex (see Appendix A.1). Clearly, the SOS index of a form is related to the positive definiteness of its SMR matrices. This is explained in the following result.

Lemma 1.2. *Let $H + L(\alpha)$ be a complete SMR matrix of $h \in \Xi_{n,2m}$. Then,*

$$\begin{aligned} \lambda(h) > 0 &\iff \exists \alpha : H + L(\alpha) > 0 \\ \lambda(h) \geq 0 &\iff \exists \alpha : H + L(\alpha) \geq 0. \end{aligned}$$

Lemma 1.2 provides necessary and sufficient conditions for establishing whether a form admits either positive definite or positive semidefinite SMR matrices in terms of its SOS index. Hence, the SOS index can be used to establish whether a form is SOS. The following result is a direct consequence of Theorem 1.4 and Lemma 1.2.

Corollary 1.2. *Let $h \in \Xi_{n,2m}$. Then, $h(x)$ is SOS if and only if*

$$\lambda(h) \geq 0. \tag{1.30}$$

Example 1.4. Let us consider $h(x)$ in (1.20). Then, by solving the EVP (1.29) with H and $L(\alpha)$ as in (1.21) we find

$$\lambda(h) = 0.0352$$

which implies that $h(x)$ is SOS. In particular, the optimal value of α in (1.29) is $\alpha^* = 0.8008$, and the corresponding decomposition of $h(x)$ in (1.27) is given by

$$\begin{aligned} h_1(x) &= x_1^2 + x_1 x_2 - 0.8008 x_2^2 \\ h_2(x) &= 0.7756 x_1 x_2 + 1.0325 x_2^2 \\ h_3(x) &= 0.5411 x_2^2. \end{aligned}$$

Example 1.5. Let us consider $h(x)$ in (1.22). This form is obviously SOS. In particular, a positive semidefinite SMR matrix of $h(x)$ is the matrix H in (1.23). It is interesting to observe that the SOS index for this form is strictly positive

$$\lambda(h) = 0.7639,$$

corresponding to the maximizer $\bar{\alpha} = (0.382, 0, 0.382, 0, 0, 0.382)'$ in (1.29). This means that $h(x)$ admits a positive definite SMR matrix. The corresponding decomposition of $h(x)$ in (1.27) is given by

$$\begin{aligned} h_1(x) &= x_1^2 - 0.3820 x_2^2 - 0.3820 x_3^2 & h_2(x) &= 0.8740 x_1 x_2 \\ h_3(x) &= 0.8740 x_1 x_3 & h_4(x) &= 1.3617 x_2^2 - 0.3877 x_3^2 \\ h_5(x) &= 0.8740 x_2 x_3 & h_6(x) &= 1.6443 x_3^2. \end{aligned}$$

For parametrized forms, the SOS index is defined as follows. Let $h(x;p)$ be a form of degree $2m$ in $x \in \mathbb{R}^n$, for any fixed parameter p. Then, the SOS index of $h(x;p)$ is the function of p

$$\lambda(h(\cdot;p)) = \max_{t,\alpha} t$$
$$\text{s.t. } H(p) + L(\alpha) - t I_{\sigma(n,m)} \geq 0$$

where the complete SMR matrix $H(p) + L(\alpha)$ is computed by considering $h(x; p)$ as a form in x with coefficients depending on p.

1.5 Matrix Forms

In this section, we introduce matrix forms and their representation.

Definition 1.8 (Matrix Form). The function $M : \mathbb{R}^n \to \mathbb{R}^{r \times r}$ is a *matrix form* of degree d in n scalar variables if

$$M_{i,j} \in \Xi_{n,d} \quad \forall i, j = 1, \dots, r. \tag{1.31}$$

We denote the set of $r \times r$ matrix forms of degree d in n scalar variables as

$$\Xi_{n,d,r}^{\sharp} = \left\{ M : \mathbb{R}^n \to \mathbb{R}^{r \times r} : (1.31) \text{ holds} \right\} \tag{1.32}$$

and the set of symmetric matrix forms as

$$\Xi_{n,d,r} = \left\{ M \in \Xi_{n,d,r}^{\sharp} : M(x) = M(x)' \quad \forall x \in \mathbb{R}^n \right\}. \tag{1.33}$$

Similarly to what has been done for forms in Definition 1.3, a matrix form $M \in \Xi_{n,d,r}^{\sharp}$ can be written as

$$M(x) = G \left(x^{\{d\}} \otimes I_r \right) \tag{1.34}$$

where $G \in \mathbb{R}^{r\sigma(n,d) \times r}$ is a suitable coefficient matrix.

Example 1.6. Consider the matrix form

$$M(x) = \begin{pmatrix} x_1^3 + 3x_1^2 x_2 & 0 \\ -x_1 x_2^2 & 2x_2^3 \end{pmatrix}.$$

One has $M \in \Xi_{2,3,2}^{\sharp}$, i.e. $n = 2$, $d = 3$ and $r = 2$. Then, $M(x)$ can be written as in (1.34) with

$$G = \begin{pmatrix} 1 & 0 & 3 & 0 & 0 & 0 & 0 & 0 \\ 0 & 0 & 0 & 0 & -1 & 0 & 0 & 2 \end{pmatrix}, \quad x^{\{d\}} = \begin{pmatrix} x_1^3 \\ x_1^2 x_2 \\ x_1 x_2^2 \\ x_2^3 \end{pmatrix}.$$

1.5.1 SMR of Matrix Forms

Matrix forms of even degree (i.e., having all elements of even degree) can be represented by introducing a suitable extension of the SMR formalism. Although this can be done for any matrix form, in the sequel the treatment will be focused on symmetric matrix forms. Let us introduce the notation

$$\Phi\left(H, x^{\{m\}}, r\right) = \left(x^{\{m\}} \otimes I_r\right)' H \left(x^{\{m\}} \otimes I_r\right). \tag{1.35}$$

Definition 1.9 (SMR of Matrix Form). Let $M \in \Xi_{n,2m,r}$ and $H \in \mathbb{S}^{r\sigma(n,m)}$ be such that

$$M(x) = \Phi\left(H, x^{\{m\}}, r\right). \tag{1.36}$$

Then, (1.36) is called a *SMR* of $M(x)$ with respect to $x^{\{m\}} \otimes I_r$. Moreover, H is called a *SMR matrix* of $M(x)$ with respect to $x^{\{m\}} \otimes I_r$.

Let us define the set of matrices H satisfying (1.36) as

$$\mathcal{H}(M) = \left\{H \in \mathbb{S}^{r\sigma(n,m)} : (1.36) \text{ holds}\right\}. \tag{1.37}$$

The following result characterizes $\mathcal{H}(M)$.

Theorem 1.5. *Let $M \in \Xi_{n,2m,r}$. Then, $\mathcal{H}(M)$ is an affine space. Moreover,*

$$\mathcal{H}(M) = \left\{H + L : H \in \mathbb{S}^{r\sigma(n,m)} \text{ satisfies (1.36)}, L \in \mathcal{L}_{n,m,r}\right\} \tag{1.38}$$

where $\mathcal{L}_{n,m,r}$ is the linear space

$$\mathcal{L}_{n,m,r} = \left\{L \in \mathbb{S}^{r\sigma(n,m)} : \Phi\left(L, x^{\{m\}}, r\right) = 0_{r \times r} \forall x \in \mathbb{R}^n\right\} \tag{1.39}$$

whose dimension is given by

$$\omega(n,m,r) = \frac{1}{2}r\left(\sigma(n,m)(r\sigma(n,m)+1) - (r+1)\sigma(n,2m)\right). \tag{1.40}$$

Proof. As in the proof of Lemma 1.1, $\mathcal{H}(M)$ in (1.37) is an affine space because $a_1 H_1 + a_2 H_2 \in \mathcal{H}(M)$ for all $H_1, H_2 \in \mathcal{H}(M)$ and for all $a_1, a_2 \in \mathbb{R}$ such that $a_1 + a_2 = 1$. This implies that $\mathcal{H}(M)$ can be written as in (1.38). Now, observe that $r\sigma(n,m)(r\sigma(n,m)+1)/2$ is the number of distinct entries of a symmetric matrix of dimension $r\sigma(n,m) \times r\sigma(n,m)$, while $r(r+1)\sigma(n,2m)/2$ is the total number of monomials in the distinct entries of a symmetric matrix form with size $r \times r$ of degree $2m$ in n scalar variables. The constraints obtained by annihilating these monomials are linear and independent, according to the same reasoning adopted in the proof of Theorem 1.2. Therefore, the dimension of $\mathcal{L}_{n,m,r}$ is $\omega(n,m,r)$. $\square$

It can be verified that $\mathscr{L}_{n,m,1}$ coincides with $\mathscr{L}_{n,m}$. Indeed, one has $\omega(n,m,1) = \omega(n,m)$. Tables 1.2 and 1.3 show the quantities $r\sigma(n,m)$ and $\omega(n,m,r)$ for some values of n,m,r.

Table 1.2 Quantity $r\sigma(n,m)$ for some values of n,m,r: (a) $r=2$; (b) $r=3$

	(a)					(b)			
	$m=1$	$m=2$	$m=3$	$m=4$		$m=1$	$m=2$	$m=3$	$m=4$
$n=1$	2	2	2	2	$n=1$	3	3	3	3
$n=2$	4	6	8	10	$n=2$	6	9	12	15
$n=3$	6	12	20	30	$n=3$	9	18	30	45
$n=4$	8	20	40	70	$n=4$	12	30	60	105

Table 1.3 Quantity $\omega(n,m,r)$ for some values of n,m,r: (a) $r=2$; (b) $r=3$

	(a)					(b)			
	$m=1$	$m=2$	$m=3$	$m=4$		$m=1$	$m=2$	$m=3$	$m=4$
$n=1$	0	0	0	0	$n=1$	0	0	0	0
$n=2$	1	6	15	28	$n=2$	3	15	36	66
$n=3$	3	33	126	330	$n=3$	9	81	297	765
$n=4$	6	105	568	1990	$n=4$	18	255	1326	4575

Let $L(\alpha)$ be a linear parametrization of $\mathscr{L}_{n,m,r}$, where $\alpha \in \mathbb{R}^{\omega(n,m,r)}$. The complete SMR of matrix forms is defined as follows.

Definition 1.10 (Complete SMR of Matrix Form). Consider any $M \in \Xi_{n,2m,r}$. Let $H \in \mathbb{S}^{r\sigma(n,m)}$ be such that (1.36) holds, and $L(\alpha)$ be a linear parametrization of $\mathscr{L}_{n,m,r}$ with $\alpha \in \mathbb{R}^{\omega(n,m,r)}$. Then, $M(x)$ can be written as

$$M(x) = \Phi\left(H + L(\alpha), x^{\{m\}}, r\right). \tag{1.41}$$

The expression in (1.41) is called a *complete SMR* of $M(x)$ with respect to $x^{\{m\}} \otimes I_r$. Moreover, $H + L(\alpha)$ is called a *complete SMR matrix* of $M(x)$ with respect to $x^{\{m\}}$.

Algorithms for the construction of the matrices H and $L(\alpha)$ in (1.41) are reported in Appendix B.

1.5.2   SOS Matrix Forms

In the sequel, it will be shown how the SOS property introduced in Section 1.4 for scalar forms can be extended to matrix forms, and how the SMR can be used to characterize this property via convex optimization.

Definition 1.11 (SOS Matrix Form). Let $M \in \Xi_{n,2m,r}$, and suppose there exist $M_1, \ldots, M_j \in \Xi_{n,m,r}^{\sharp}$ for some integer $j \geq 1$ such that

$$M(x) = \sum_{i=1}^{j} M_i(x)' M_i(x). \tag{1.42}$$

Then, $M(x)$ is said to be a *SOS matrix form*.

   We denote the set of SOS matrix forms as

$$\Sigma_{n,2m,r} = \{M \in \Xi_{n,2m,r} : M(x) \text{ is SOS}\}. \tag{1.43}$$

The SMR directly allows one to establish whether a matrix form is SOS or not via an LMI feasibility test as described in the next result.

Theorem 1.6. *Let $H + L(\alpha)$ be a complete SMR matrix of $M \in \Xi_{n,2m,r}$. Then, $M(x)$ is SOS if and only if*

$$\exists \alpha : H + L(\alpha) \geq 0. \tag{1.44}$$

Proof. Analogous to the proof of Theorem 1.4.   □

   For any $M \in \Xi_{n,2m,r}$ we define the SOS index via the solution of an EVP similarly to (1.29) as follows.

Definition 1.12 (SOS Index of Matrix Form). Let $M \in \Xi_{n,2m,r}$ and define the EVP

$$\begin{aligned} \lambda(M) = \max_{t,\alpha} \ & t \\ \text{s.t. } & H + L(\alpha) - t I_{r\sigma(n,m)} \geq 0 \end{aligned} \tag{1.45}$$

where $H + L(\alpha)$ is a complete SMR matrix of $M(x)$. Then, $\lambda(M)$ is called *SOS index of $M(x)$*.

Corollary 1.3. *Let $M \in \Xi_{n,2m,r}$. Then, $M(x)$ is SOS if and only if*

$$\lambda(M) \geq 0. \tag{1.46}$$

   Corollary 1.3 states that one can establish whether a matrix form $M(x)$ is SOS via the condition (1.46), which requires the computation of the SOS index (1.45).

Example 1.7. Consider the matrix form

$$M(x) = \begin{pmatrix} x_1^2 + x_2^2 & -x_1x_2 + x_2^2 \\ \star & x_1^2 + 2x_2^2 \end{pmatrix}. \tag{1.47}$$

One has that $M \in \Xi_{2,2,2}$, i.e. $n = 2$, $m = 1$ and $r = 2$. Then, $M(x)$ can be written as in (1.41) with

$$x^{\{m\}} = x, \quad H = \begin{pmatrix} 1 & 0 & 0 & -0.5 \\ \star & 1 & -0.5 & 0 \\ \star & \star & 1 & 1 \\ \star & \star & \star & 2 \end{pmatrix}, \quad L(\alpha) = \begin{pmatrix} 0 & 0 & 0 & -\alpha_1 \\ \star & 0 & \alpha_1 & 0 \\ \star & \star & 0 & 0 \\ \star & \star & \star & 0 \end{pmatrix}. \tag{1.48}$$

In fact, the dimension of α is given by $\omega(2,1,2) = 1$, according to (1.40). We compute the SOS index of $M(x)$ in (1.45), and find that $\lambda(M) = 0.1335$ (the optimal value of α in (1.45) is $\alpha^* = 0.1536$). Hence, $M(x)$ is SOS.

Example 1.8. Consider the matrix form

$$M(x) = \begin{pmatrix} x_1^4 & x_1x_2^3 \\ \star & x_2^4 \end{pmatrix}. \tag{1.49}$$

One has that $M \in \Xi_{2,4,2}$, i.e. $n = 2$, $m = 2$ and $r = 2$. Then, $M(x)$ can be written as in (1.41) with

$$x^{\{m\}} = \begin{pmatrix} x_1^2 \\ x_1x_2 \\ x_2^2 \end{pmatrix}, \quad H = \begin{pmatrix} 1 & 0 & 0 & 0 & 0 & 0 \\ \star & 0 & 0 & 0 & 0 & 0 \\ \star & \star & 0 & 0 & 0 & 1 \\ \star & \star & \star & 0 & 0 & 0 \\ \star & \star & \star & \star & 0 & 0 \\ \star & \star & \star & \star & \star & 1 \end{pmatrix}$$

$$L(\alpha) = \begin{pmatrix} 0 & 0 & 0 & -\alpha_1 & -\alpha_3 & -\alpha_2 - \alpha_4 \\ \star & 0 & \alpha_1 & 0 & \alpha_2 & -\alpha_5 \\ \star & \star & 2\alpha_3 & \alpha_4 & 0 & -\alpha_6 \\ \star & \star & \star & 2\alpha_5 & \alpha_6 & 0 \\ \star & \star & \star & \star & 0 & 0 \\ \star & \star & \star & \star & \star & 0 \end{pmatrix}. \tag{1.50}$$

The dimension of α is $\omega(2,2,2) = 6$. For this matrix form we find that $\lambda(M) = -0.1202$, which implies that $M(x)$ is not SOS.

1.6   Positive Forms

This section illustrates how the SMR can be used for establishing positivity of forms.

1.6.1   Positivity Index

Let us start by introducing the following definition.

Definition 1.13 (Positive Definite Form). A form $h \in \Xi_{n,2m}$ is *positive definite* (resp., *semidefinite*) if $h(x) > 0$ (resp., $h(x) \geq 0$) for all $x \neq 0_n$.

We denote the set of positive semidefinite forms of degree $2m$ in n scalar variables by

$$\Omega_{n,2m} = \{h \in \Xi_{n,2m} : h(x) \text{ is positive semidefinite}\}. \tag{1.51}$$

Positive definiteness of forms can be also expressed via the following index.

Definition 1.14 (Positivity Index). Let $h \in \Xi_{n,2m}$ and define

$$\mu(h) = \min_{x \in \mathscr{C}_{n,m}} h(x) \tag{1.52}$$

where

$$\mathscr{C}_{n,m} = \left\{x \in \mathbb{R}^n : \left\|x^{\{m\}}\right\| = 1\right\}. \tag{1.53}$$

Then, $\mu(h)$ is called *positivity index* of $h(x)$.

Let us observe that, depending on the chosen power vector $x^{\{m\}}$, one can obtain various shapes for the set $\mathscr{C}_{n,m}$ in (1.53). In particular, if $x^{\{m\}}$ is chosen to satisfy (1.9), then one has

$$\mathscr{C}_{n,m} = \{x \in \mathbb{R}^n : \|x\| = 1\}.$$

For parametrized forms, the positivity index is defined as follows. Let $h(x; p)$ be a form of degree $2m$ in $x \in \mathbb{R}^n$ for any fixed parameter p. Then, the positivity index of $h(x; p)$ is the function of p

$$\mu(h(\cdot; p)) = \min_{x \in \mathscr{C}_{n,m}} h(x; p).$$

The following result shows that positive definiteness and semidefiniteness of a form $h(x)$ can be directly established from the positivity index $\mu(h)$.

Theorem 1.7. *Let $h \in \Xi_{n,2m}$. Then,*

$$h(x) \text{ is positive definite} \iff \mu(h) > 0$$
$$h(x) \text{ is positive semidefinite} \iff \mu(h) \geq 0.$$

Proof. Let us observe that $\mathscr{C}_{n,m}$ in (1.53) is a closed hyper-surface in $\mathbb{R}^n$ centered in 0_n, and that $x^{\{m\}}$ is a vector of forms of degree m. For any $x \in \mathbb{R}_0^n$, let $a(x) = \left\|x^{\{m\}}\right\|^{-\frac{1}{m}}$; then one has that $a(x)x \in \mathscr{C}_{n,m}$. Moreover, since $h(x)$ is a form of degree $2m$, one has that

$$h(a(x)x) = a(x)^{2m}h(x)$$
$$= \left\|x^{\{m\}}\right\|^{-2} h(x)$$

which implies that, for all $x \in \mathbb{R}_0^n$,

$$h(x) > 0 \iff h(a(x)x) > 0$$
$$h(x) \geq 0 \iff h(a(x)x) \geq 0$$

and hence the theorem holds. $\square$

A property of positive forms that will be useful in the following is given next.

Theorem 1.8. *Let $h \in \Xi_{n,2m}$ and assume that $h(x)$ is positive definite. Then, there exists $\varepsilon > 0$ such that the form $h(x) - \varepsilon\|x\|^{2m}$ is positive definite.*

Proof. By Theorem 1.7, one has $\mu(h) > 0$. Let us define

$$\delta = \max_{x \in \mathscr{C}_{n,2m}} \|x\|^{2m}. \tag{1.54}$$

Being $\mathscr{C}_{n,2m}$ compact, δ is finite and positive. Now, consider the form $h_1(x) = h(x) - \varepsilon\|x\|^{2m}$. One has

$$\mu(h_1) = \min_{x \in \mathscr{C}_{n,2m}} h(x) - \varepsilon\|x\|^{2m}$$
$$\geq \min_{x \in \mathscr{C}_{n,2m}} h(x) - \varepsilon \max_{x \in \mathscr{C}_{n,2m}} \|x\|^{2m}$$
$$= \mu(h) - \varepsilon\delta.$$

Then, by choosing $\varepsilon < \delta^{-1}\mu(h)$, one has $\mu(h_1) > 0$, and therefore $h_1(x)$ is positive definite. $\square$

1.6.2   Sufficient Condition for Positivity

The SMR can be used to investigate whether a form is either positive definite or positive semidefinite. Indeed, the following result holds.

Theorem 1.9. *Let $h \in \Xi_{n,2m}$. Then,*

$$\lambda(h) \leq \mu(h). \tag{1.55}$$

Proof. From (1.29) one has that

$$H(\bar{\alpha}) - \lambda(h)I_{\sigma(n,m)} \geq 0 \tag{1.56}$$

for some $\bar{\alpha}$. By pre- and post-multiplying by $x^{\{m\}'}$ and $x^{\{m\}}$ respectively, the relation (1.56) turns into

$$h(x) - \lambda(h)x^{\{m\}'}x^{\{m\}} \geq 0 \quad \forall x \in \mathbb{R}^n.$$

Let us consider now this property for $x \in \mathscr{C}_{n,m}$. One has that:

$$h(x) - \lambda(h) \geq 0 \quad \forall x \in \mathscr{C}_{n,m}.$$

In particular,

$$\min_{x \in \mathscr{C}_{n,m}} h(x) - \lambda(h) \geq 0$$

which is equivalent to (1.55), and hence the theorem holds.   □

Theorem 1.9 states that a lower bound to the positivity index $\mu(h)$ is given by the SOS index $\lambda(h)$ in (1.29). The following result is a direct application of Theorems 1.7 and 1.9.

Corollary 1.4. *Let $h \in \Xi_{n,2m}$. Then:*

$$\lambda(h) > 0 \quad \Rightarrow \quad h(x) \text{ is positive definite}$$
$$\lambda(h) \geq 0 \quad \Rightarrow \quad h(x) \text{ is positive semidefinite.}$$

Corollary 1.4 provides sufficient conditions to check whether a form is either positive definite or positive semidefinite, which require to solve a convex optimization for computing $\lambda(h)$. The next chapter investigates the necessity of this condition.

Example 1.9. Let us consider the forms in Examples 1.2 and 1.3. In Examples 1.4 and 1.5 it has been found that $\lambda(h) > 0$ for these forms, hence Corollary 1.4 implies that these forms are positive definite.

1.6.3 Positive Matrix Forms

In this section we extend the previous results to the case of matrix forms. First of all, let us introduce the following definition.

Definition 1.15 (Positive Definite Matrix Form). A matrix form $M \in \Xi_{n,2m,r}$ is *positive definite* (resp., *semidefinite*) if $M(x) > 0$ (resp., $M(x) \geq 0$) for all $x \neq 0_n$.

Similarly to the case of scalar forms, positive definiteness of matrix forms can be expressed via an appropriate index.

Definition 1.16 (Positivity Index for Matrix Form). Let $M \in \Xi_{n,2m,r}$ and define

$$\mu(M) = \min_{x \in \mathscr{C}_{n,m}} \lambda_{min}(M(x)). \tag{1.57}$$

Then, $\mu(M)$ is called *positivity index* of $M(x)$.

The following result explains how positive definiteness and semidefiniteness of a matrix form $M(x)$ can be directly established from $\mu(M)$.

Theorem 1.10. *Let $M \in \Xi_{n,2m,r}$. Then,*

$$M(x) \text{ is positive definite} \iff \mu(M) > 0$$
$$M(x) \text{ is positive semidefinite} \iff \mu(M) \geq 0.$$

Proof. Analogous to the proof of Theorem 1.7.   □

The following result states that a lower bound to $\mu(M)$ can be obtained by using the SMR.

Theorem 1.11. *Let $M \in \Xi_{n,2m,r}$. Then,*

$$\lambda(M) \leq \mu(M). \tag{1.58}$$

Moreover,
$$\lambda(M) > 0 \implies M(x) \text{ is positive definite}$$
$$\lambda(M) \geq 0 \implies M(x) \text{ is positive semidefinite}.$$

Proof. The proof of (1.58) is analogous to the proof of Theorem 1.9. Then, the conditions for positive definiteness or semidefiniteness of $M(x)$ via its SOS index follow from (1.58) and Theorem 1.10.   □

Example 1.10. Let us consider the matrix forms in Examples 1.7 and 1.8. The SOS index for the matrix form in (1.47) is greater than 0, hence implying by Theorem 1.11 that $M(x)$ is positive definite. On the contrary, for the matrix form in (1.49) one has an SOS index lesser than 0 and hence positivity cannot be established. Indeed, for

$$x = \begin{pmatrix} 0.5104 \\ 0.9187 \end{pmatrix}$$

it turns out that the minimum eigenvalue of $M(x)$ is equal to -0.1202.

1.7   Positive Polynomials on Ellipsoids

In this section we describe a method for investigating positivity of a polynomial over an ellipsoid. Let $f(x)$ be a polynomial in $x \in \mathbb{R}^n$. We consider the problem of establishing whether

$$f(x) > 0 \quad \forall x \in \mathscr{B}(Q,c) \tag{1.59}$$

where $\mathscr{B}(Q,c)$ denotes the ellipsoid

$$\mathscr{B}(Q,c) = \{x \in \mathbb{R}^n : x'Qx = c\} \tag{1.60}$$

for some $c \in \mathbb{R}$, $c \geq 0$, and $Q \in \mathbb{S}^n$, $Q > 0$. We refer to Q as the *shape matrix* of the ellipsoid.

The strategy we will present relies on casting problem (1.59) into the problem of establishing whether a suitable form is positive definite.

1.7.1   Solution via Positivity Test on a Form

Let us first consider the case in which $f(x)$ is a generic polynomial of degree m. Without loss of generality, we assume that

$$f(\bar{x}) > 0, \quad \bar{x} = \sqrt{\frac{c}{Q_{1,1}}}(1,0,\ldots,0)'. \tag{1.61}$$

Clearly, if this is not the case, $f(x)$ cannot be positive on the ellipsoid $\mathscr{B}(Q,c)$. The following result provides an alternative to (1.59) based on the positivity of a suitable symmetric polynomial.

Theorem 1.12. *Condition* (1.59) *holds if and only if*

$$f(x)f(-x) > 0 \quad \forall x \in \mathscr{B}(Q,c). \tag{1.62}$$

Proof. (Necessity) Let us assume that (1.59) holds, and suppose by contradiction that $\exists y \in \mathscr{B}(Q,c)$ such that $f(y)f(-y) \leq 0$. Since $f(y) > 0$ from (1.59), this implies that also $f(-y) \leq 0$. However, since $-y \in \mathscr{B}(Q,c)$, this contradicts (1.59).

(Sufficiency) Let us assume that (1.62) holds, and suppose by contradiction that $\exists y \in \mathscr{B}(Q,c)$ such that $f(y) \leq 0$. Let us first consider the case $f(y) = 0$. This is not possible because it implies $f(y)f(-y) = 0$, which contradicts (1.62). Hence, let us consider the case $f(y) < 0$. Due to (1.61), being $\mathscr{B}(Q,c)$ connected and $f(x)$ a continuous function, it follows that

$$\exists z \in \mathscr{B}(Q,c) : f(z) = 0.$$

But this implies $f(z)f(-z) = 0$, which contradicts (1.62). $\qquad\square$

Let us observe that the polynomial $f(x)f(-x)$ is symmetric with respect to the origin, and hence can be written as

$$f(x)f(-x) = \sum_{i=0}^{m} \bar{f}_{2i}(x) \qquad (1.63)$$

for suitable forms $\bar{f}_{2i}(x) \in \Xi_{n,2i}$, $i = 0, \ldots, m$. Let us introduce the function

$$w(x;c) = \sum_{i=0}^{m} \bar{f}_{2i}(x) \left(\frac{x'Qx}{c} \right)^{m-i} \qquad (1.64)$$

Notice that
$$w(x;c) = f(x)f(-x) \quad \forall x \in \mathcal{B}(Q,c).$$

Moreover, for any fixed c, $w(x;c)$ is a form in x of degree $2m$, i.e. $w(\cdot;c) \in \Xi_{n,2m}$ is a parametrized form.

The following result states that the positivity of the polynomial $f(x)f(-x)$ over the ellipsoid $\mathcal{B}(Q,c)$ is equivalent to the positive definiteness of $w(x;c)$.

Theorem 1.13. *Condition* (1.62) *holds if and only if*

$$w(x;c) > 0 \quad \forall x \neq 0_n. \qquad (1.65)$$

Proof. (Necessity) Let us assume that (1.62) holds, and suppose by contradiction that $\exists y \in \mathbb{R}_0^n$ such that $w(y;c) \leq 0$. Let us define $z = \beta y$, where $\beta = \sqrt{c(y'Qy)^{-1}}$. Then, $z \in \mathcal{B}(Q,c)$. Moreover,

$$w(z;c) = \beta^{2m} w(y;c)$$

because $w(x;c)$ is a form of degree $2m$ in x. Since $w(y;c) \leq 0$, one also has $w(z;c) \leq 0$. Moreover, $w(z;c) = f(z)f(-z)$ because $z \in \mathcal{B}(Q,c)$. Therefore,

$$\exists z \in \mathcal{B}(Q,c) : \ f(z)f(-z) \leq 0$$

which contradicts (1.62).

(Sufficiency) Let us suppose that (1.65) holds. This implies that (1.62) holds since $w(x;c) = f(x)f(-x)$ for all $x \in \mathcal{B}(Q,c)$. $\qquad\square$

By exploiting Theorems 1.12 and 1.13, one obtains the following result, which allows one to formulate problem (1.59) as a positivity test on a form.

Theorem 1.14. *Condition* (1.59) *holds if and only if* (1.65) *holds.*

1.7.2   Even Polynomials

The condition provided by Theorem 1.14 can be simplified if $f(x)$ satisfies the following definition.

Definition 1.17 (Even Polynomial). The function $f : \mathbb{R}^n \to \mathbb{R}$ is an *even polynomial* of degree less than or equal to $2m$ in n scalar variables if

$$f(x) = \sum_{i=0}^{m} f_{2i}(x) \tag{1.66}$$

for some $f_{2i}(x) \in \Xi_{n,2i}$, $i = 0, \ldots, m$.

Hence, let us suppose that $f(x)$ is an even polynomial of degree $2m$ as in (1.66), and let us define $w(x;c)$ as

$$w(x;c) = \sum_{i=0}^{m} f_{2i}(x) \left(\frac{x'Qx}{c} \right)^{m-i}. \tag{1.67}$$

The following result generalizes Theorem 1.14 by including its extension to the case of even polynomials.

Theorem 1.15. *Condition* (1.59) *holds if and only if one of the following conditions holds:*

1. $f(x)$ *is a polynomial of degree m and* (1.65) *holds with $w(x;c)$ as in* (1.63)–(1.64);
2. $f(x)$ *is an even polynomial of degree $2m$ and* (1.65) *holds with $w(x;c)$ as in* (1.66)–(1.67).

Proof. Item 1 is just Theorem 1.14. Turning to item 2, let $w(x;c)$ be given by (1.67). Since $w(x;c) \in \Xi_{n,2m}$ one has that $w(y;c) = \beta^{2m} w(x;c)$ whenever $y = \beta x$. With $\beta = \sqrt{c(x'Qx)^{-1}}$ one has that $y \in \mathcal{B}(Q,c)$ and, hence, $w(y;c) = f(y)$. Taking into account that $\beta > 0$, it follows that (1.59) and (1.65) are equivalent.   □

Summarizing, the conditions proposed by Theorem 1.15 rely on a suitable homogenization of the polynomial $f(x)$ and establish that (1.59) is equivalent to the positivity of the form $w(x;c)$. Notice that $w(x;c)$ is constructed by direct homogenization of $f(x)$ if this is an even polynomial, while it comes from homogenization of $f(x)f(-x)$ in the general case (thus requiring to double the degree of the form $w(x;c)$ with respect to that of $f(x)$). A sufficient condition for positivity of a polynomial over an ellipsoid can be formulated in terms of the SOS index of the parametrized form $w(x;c)$.

Theorem 1.16. *Condition* (1.59) *holds if one of the following conditions holds:*

1. $f(x)$ *is a polynomial of degree m and $\lambda(w(\cdot;c)) > 0$ with $w(x;c)$ as in* (1.63)–(1.64);

2. $f(x)$ is an even polynomial of degree $2m$ and $\lambda(w(\cdot;c)) > 0$ with $w(x;c)$ as in (1.66)–(1.67).

Proof. It directly follows from Theorem 1.15 and Corollary 1.4. □

Example 1.11. Let us consider the problem to establish whether (1.59) holds with

$$f(x) = 2 + x_1 + x_2^2$$
$$Q = I_2$$
$$c = 1.$$

We have $n = 2$ and $m = 2$. By applying (1.63)–(1.64), we get

$$w(x;c) = 4\frac{(x_1^2 + x_2^2)^2}{c^2} + (4x_2^2 - x_1^2)\frac{x_1^2 + x_2^2}{c} + x_2^4.$$

Then, we find that $\lambda(w(\cdot;c)) = 3.0000$, which implies from Theorem 1.16 that (1.59) holds.

1.8 Positive Matrix Polynomials on the Simplex

In this section we describe a method for investigating positive definiteness of a matrix of polynomials on the positive octant. Let us introduce the following definitions.

Definition 1.18 (Simplex). The set defined as

$$\Upsilon_n = \left\{ x \in \mathbb{R}^n : \sum_{i=1}^n x_i = 1, \ x_i \geq 0 \right\} \tag{1.68}$$

is called *simplex*.

Definition 1.19 (Matrix Polynomial). The function $N : \mathbb{R}^n \to \mathbb{R}^{r \times r}$ is a *matrix polynomial* of degree less than or equal to m in n scalar variables if

$$N(x) = \sum_{i=0}^m N_i(x) \tag{1.69}$$

with $N_i(x) \in \Xi_{n,i,r}^\sharp$, $i = 0, \ldots, m$. Moreover, if $N(x) = N(x)'$, then $N(x)$ is a *symmetric matrix polynomial*.

The problem considered in this section is formulated as follows: establish whether a symmetric matrix polynomial $N(x)$ satisfies the condition

$$N(x) > 0 \quad \forall x \in \Upsilon_n. \tag{1.70}$$

In order to address (1.70), let us define

$$M(x) = \sum_{i=1}^{m} N_i(x) \left(\sum_{j=1}^{n} x_j \right)^{m-i}. \tag{1.71}$$

It is straightforward to verify that $M \in \Xi_{n,m,r}$, i.e. $M(x)$ is a matrix form. By observing that $N(x) = M(x)$ for all $x \in \Upsilon_n$, condition (1.70) can be reformulated as follows.

Lemma 1.3. *Condition* (1.70) *holds if and only if*

$$M(x) > 0 \quad \forall x \in \Upsilon_n. \tag{1.72}$$

Hence, without loss of generality we can consider the problem of establishing whether (1.72) holds.

Theorem 1.17. *Let $M \in \Xi_{n,m,r}$. Then, condition* (1.72) *holds if and only if*

$$M(\mathrm{sq}(x)) > 0 \quad \forall x \in \mathbb{R}_0^n. \tag{1.73}$$

Proof. (Necessity) Let us assume that (1.72) holds, and suppose by contradiction that $\exists y \in \mathbb{R}_0^n$ such that $M(\mathrm{sq}(y)) \not> 0$. Let us define

$$z = \beta \, \mathrm{sq}(y)$$

$$\beta = \left(\sum_{i=1}^{n} y_i^2 \right)^{-1}.$$

It follows that $z \in \Upsilon_n$. Moreover,

$$M(z) = \beta^m M(\mathrm{sq}(y))$$

because $M(x)$ is a matrix form of degree m in x. Since $M(\mathrm{sq}(y)) \not> 0$, then also $M(z) \not> 0$. Therefore, there exists $z \in \Upsilon_n$ such that $M(z) \not> 0$, thus contradicting (1.72).

(Sufficiency) Let us assume that (1.73) holds, and suppose by contradiction that $\exists y \in \Upsilon_n$ such that $M(y) \not> 0$. Define $z = \mathrm{sqr}(y)$. Then, $M(\mathrm{sq}(z)) = M(y)$. Therefore, there exists $z \neq 0_n$ such that $M(\mathrm{sq}(z)) \not> 0$, thus contradicting (1.73). □

By using the SOS index, it is possible to formulate a sufficient condition for (1.72) through a convex optimization problem. This is explained in the following result.

Theorem 1.18. *Let $M \in \Xi_{n,m,r}$ and define $M_1(x) = M(\text{sq}(x))$. Let us suppose that*

$$\lambda(M_1) > 0. \tag{1.74}$$

Then, condition (1.72) holds.

Proof. Let us suppose that (1.74) holds. This means that $M(\text{sq}(x))$ is positive definite from Theorem 1.11. Therefore, (1.73) holds, hence implying (1.72).   □

Example 1.12. Let us consider the problem of establishing whether (1.70) holds for

$$N(x) = \begin{pmatrix} x_1 + 1 & x_1 \\ \star & x_2 + 1 \end{pmatrix}.$$

Then, $M(x)$ in (1.71) is given by

$$M(x) = \begin{pmatrix} 2x_1 + x_2 & x_1 \\ \star & x_1 + 2x_2 \end{pmatrix}.$$

It turns out that $\lambda(M_1) = 0.3820$. Therefore, we conclude from Theorem 1.18 that (1.70) holds.

1.9   Extracting Power Vectors from Linear Spaces

This section investigates the problem of finding power vectors in a linear space, i.e. determine $x \in \mathbb{R}^n$ such that

$$x^{\{m\}} \in \mathcal{V} \tag{1.75}$$

where $\mathcal{V} \subseteq \mathbb{R}^n$ is a given linear space. As it will become clear in the next chapters, solving this problem is important in order to establish non-conservatism and optimality of several robustness analysis problems.

First of all, let us observe that $x = 0_n$ always satisfies (1.75), and hence this trivial solution can be neglected. Moreover, we observe that

$$x^{\{m\}} \in \mathcal{V} \implies (\beta x)^{\{m\}} \in \mathcal{V} \quad \forall \beta \in \mathbb{R}$$

since $x^{\{m\}}$ is homogeneous and $\mathcal{V}$ is a linear space, and hence the search can be restricted to normalized values of x. Therefore, the problem boils down to finding the set

$$\mathcal{X} = \left\{ x \in \mathbb{R}^n : \|x\| = 1, \, x^{\{m\}} \in \text{img}(V) \right\} \tag{1.76}$$

where $V \in \mathbb{R}^{\sigma(n,m) \times u}$ is a given matrix satisfying $\mathcal{V} = \text{img}(V)$ and

$$\text{rank}(V) = u = \dim(\mathcal{V}). \tag{1.77}$$

In the sequel we will adopt the notation

$$V = (v_1, \ldots, v_u) \tag{1.78}$$

where $v_1, \ldots, v_u \in \mathbb{R}^{\sigma(n,m)}$ are linearly independent vectors.

1.9.1 Basic Procedure

Let us first consider the case where the dimension u in (1.77) satisfies

$$u \leq m+1. \tag{1.79}$$

From (1.78), any $v \in \text{img}(V)$ can be written as

$$v = \sum_{i=1}^{u} s_i v_i \tag{1.80}$$

where $s \in \mathbb{R}^u$ is a suitable parameter vector. Hence, (1.75) becomes

$$x^{\{m\}} = \sum_{i=1}^{u} s_i v_i. \tag{1.81}$$

Let us select two variables x_j and x_k with $1 \leq j \leq n$, $1 \leq k \leq n$, $j \neq k$. Let us consider the rows $r_1, \ldots, r_u$ of V corresponding to the monomials $x_k^m, x_j x_k^{m-1}, \ldots, x_j^{u-1} x_k^{m-u+1}$ in $x^{\{m\}}$. By means of pivot operations, it is possible to obtain a new base for $\text{img}(V)$ by constructing a matrix $\tilde{V}$ such that the rows $\tilde{r}_i$ of $\tilde{V}$, $i = 1, \ldots, u$, satisfy

$$\begin{pmatrix} \tilde{r}_u \\ \tilde{r}_{u-1} \\ \vdots \\ \tilde{r}_1 \end{pmatrix} = I_u. \tag{1.82}$$

In this new base, every $v \in \text{img}(V)$ can be written as a linear combination of the column vectors $\tilde{v}_i$ of $\tilde{V}$, with parameter vector $\tilde{s}$. It is straightforward to verify that in the new base one has

$$\tilde{s}_i = x_j^{u-i} x_k^{m-u+i}, \ 1 \leq i \leq u$$

and hence (1.81) can be rewritten as

$$x^{\{m\}} = \sum_{i=1}^{u} x_j^{u-i} x_k^{m-u+i} \tilde{v}_i. \tag{1.83}$$

This allows one to write $x^{\{m\}}$ as a homogeneous function of the variables x_j and x_k only. In this way, by using relations between the monomials of $x^{\{m\}}$ given by rows of $\tilde{V}$ not used yet, one can write a homogeneous equation in the variables x_j and x_k that all solutions in $\mathcal{X}$ must satisfy.

In order to obtain the equation with lowest degree, one may choose the rows a and b of (1.83) corresponding to the monomials $x_h x_k^{m-1}$ and $x_j x_h x_k^{m-2}$, for some $h \neq j, h \neq k$. It turns out that

$$x_h x_k^{m-1} = \sum_{i=1}^{u} x_j^{u-i} x_k^{m-u+i} \tilde{V}_{a,i} \tag{1.84}$$

$$x_j x_h x_k^{m-2} = \sum_{i=1}^{u} x_j^{u-i} x_k^{m-u+i+1} \tilde{V}_{b,i}. \tag{1.85}$$

By multiplying (1.84) by x_j, (1.85) by x_k, and equating the right hand sides, one gets

$$x_j \left(\sum_{i=1}^{u} x_j^{u-i} x_k^{m-u+i} \tilde{V}_{a,i} \right) = \sum_{i=1}^{u} x_j^{u-i} x_k^{m-u+i+1} \tilde{V}_{b,i}. \tag{1.86}$$

This equation can be solved in the ratio $x_j x_k^{-1}$ by finding the roots of a one-variable polynomial of degree u. Indeed, one has that (1.86) is equivalent to

$$\tilde{V}_{a,1} \left(\frac{x_j}{x_k} \right)^u + \sum_{i=2}^{u} (\tilde{V}_{a,i} - \tilde{V}_{b,i-1}) \left(\frac{x_j}{x_k} \right)^{u-i+1} - \tilde{V}_{b,u} = 0. \tag{1.87}$$

Since (1.83) can be rewritten as

$$x^{\{m\}} = x_k^m \sum_{i=1}^{u} \left(\frac{x_j}{x_k} \right)^{u-i} \tilde{v}_i \tag{1.88}$$

the ratios $x_h x_k^{-1}$, $h \neq j$, can be obtained from the rows of (1.88) corresponding to monomials $x_h x_k^{m-1}$ in $x^{\{m\}}$. From the constructed x, the corresponding solution in $\mathcal{X}$ is simply obtained via the normalization $x\|x\|^{-1}$.

This procedure provides all the elements of $\mathcal{X}$ but, in principle, it may also introduce spurious solutions. The feasibility of the solution candidates can be checked by constructing the vector $\hat{x}^{\{m\}}$ for every candidate $\hat{x}$ and testing whether $\hat{x}^{\{m\}}$ belongs to img(V) or not.

If it is not possible to reach the form (1.82), i.e. the pivot procedure stops before the u-th step, the solutions of $\mathcal{X}$ can be found by considering a smaller number of vectors $\tilde{v}_i$ and, consequently, by solving an equation of lower degree.

Example 1.13. Consider the problem of computing the set $\mathcal{X}$ in (1.76), with $n = 2$, $m = 2$ and

$$x^{\{m\}} = \begin{pmatrix} x_1^2 \\ x_1 x_2 \\ x_2^2 \end{pmatrix}, \quad V = \begin{pmatrix} 5 & 2 \\ 3 & 0 \\ 2 & -1 \end{pmatrix}.$$

Observe that $u = 2$ and hence the procedure proposed above can be applied, being $u \le m+1$. Recall that the aim is to find $x \in \mathbb{R}^n$ such that (1.81) holds for some $s \in \mathbb{R}^u$.

Let us select $j = 1$ and $k = 2$. By performing the pivot procedure on matrix V with respect to monomials x_2^2 and $x_1 x_2$, one gets the matrix

$$\tilde{V} = \begin{pmatrix} 3 & -2 \\ 1 & 0 \\ 0 & 1 \end{pmatrix} \begin{array}{l} \rightarrow \\ \rightarrow \\ \rightarrow \end{array} \begin{array}{l} \text{row corresponding to} \quad x_1^2 \\ \qquad\qquad " \qquad\qquad x_1 x_2 \\ \qquad\qquad " \qquad\qquad x_2^2 \end{array}$$

whose columns form a new base for $\mathcal{V}$. In this new base, $\tilde{V}\tilde{s} = x^{\{m\}}$ is satisfied if and only if

$$\tilde{s} = \begin{pmatrix} x_1 x_2 \\ x_2^2 \end{pmatrix}.$$

From the first row, one has the equation

$$x_1^2 = 3 x_1 x_2 - 2 x_2^2$$

which can be written in the form (1.87) as

$$\left(\frac{x_1}{x_2} \right)^2 - 3 \left(\frac{x_1}{x_2} \right) + 2 = 0$$

thus yielding the solutions $x_1 = x_2$ and $x_1 = 2x_2$. By normalizing the vectors resulting from these solutions, one obtains the sought set

$$\mathcal{X} = \left\{ \frac{1}{\sqrt{2}} \begin{pmatrix} 1 \\ 1 \end{pmatrix}, \frac{1}{\sqrt{5}} \begin{pmatrix} 2 \\ 1 \end{pmatrix} \right\}.$$

It can be easily verified that both elements of $\mathcal{X}$ satisfy $x^{\{m\}} \in \text{img}(V)$ as desired.

Example 1.14. Consider the problem of computing the set $\mathcal{X}$ in (1.76), with $n = 4$, $m = 2$ and

$$x^{\{m\}} = \begin{pmatrix} x_1^2 \\ x_1 x_2 \\ x_1 x_3 \\ x_1 x_4 \\ x_2^2 \\ x_2 x_3 \\ x_2 x_4 \\ x_3^2 \\ x_3 x_4 \\ x_4^2 \end{pmatrix}, \quad V = \begin{pmatrix} 0.2405 & 0.4412 \\ 0.1771 & 0.4150 \\ -0.4999 & 0.1360 \\ 0.0977 & 0.3823 \\ 0.1198 & 0.3914 \\ -0.4915 & 0.1395 \\ 0.0481 & 0.3619 \\ -0.4020 & 0.1763 \\ -0.4810 & 0.1438 \\ -0.0139 & 0.3363 \end{pmatrix}.$$

Again, the basic procedure can be applied being $u = 2 \le m + 1 = 3$.

Let us select $j = 3$ and $k = 4$. By performing the pivot procedure on matrix V with respect to monomials x_4^2 and $x_3 x_4$ one gets

$$
\tilde{V} =
\begin{pmatrix}
-0.5446 & 1.5446 \\
-0.4089 & 1.4089 \\
1.0404 & -0.0404 \\
-0.2389 & 1.2389 \\
-0.2863 & 1.2863 \\
1.0225 & -0.0225 \\
-0.1328 & 1.1328 \\
0.8309 & 0.1691 \\
1.0000 & 0.0000 \\
0.0000 & 1.0000
\end{pmatrix}
\begin{array}{l}
\rightarrow \text{ row corresponding to } \quad x_1^2 \\
\rightarrow \qquad\qquad\qquad '' \qquad\qquad x_1 x_2 \\
\rightarrow \qquad\qquad\qquad '' \qquad\qquad x_1 x_3 \\
\rightarrow \qquad\qquad\qquad '' \qquad\qquad x_1 x_4 \\
\rightarrow \qquad\qquad\qquad '' \qquad\qquad x_2^2 \\
\rightarrow \qquad\qquad\qquad '' \qquad\qquad x_2 x_3 \\
\rightarrow \qquad\qquad\qquad '' \qquad\qquad x_2 x_4 \\
\rightarrow \qquad\qquad\qquad '' \qquad\qquad x_3^2 \\
\rightarrow \qquad\qquad\qquad '' \qquad\qquad x_3 x_4 \\
\rightarrow \qquad\qquad\qquad '' \qquad\qquad x_4^2
\end{array}
$$

and therefore $\tilde{V}\tilde{s} = x^{\{m\}}$ is satisfied if and only if

$$
\tilde{s} = \begin{pmatrix} x_3 x_4 \\ x_4^2 \end{pmatrix}.
$$

By choosing $h = 2$ and following the same procedure outlined in (1.84)–(1.87), one obtains the equation

$$
-0.1328 \left(\frac{x_3}{x_4}\right)^2 + 0.1103 \left(\frac{x_3}{x_4}\right) + 0.0225 = 0
$$

which yields the solutions $x_3 = x_4$ and $x_3 = -0.1691 x_4$. The corresponding values of x_1 and x_2 can be read directly in the fourth and seventh rows of vectors $\tilde{V}(x_3 x_4, x_4^2)'$. Indeed, for $x_3 = x_4$ one has

$$
\tilde{V} \begin{pmatrix} x_4^2 \\ x_4^2 \end{pmatrix} = x_4^2
\begin{pmatrix}
1.0000 \\
1.0000 \\
1.0000 \\
1.0000 \\
1.0000 \\
1.0000 \\
1.0000 \\
1.0000 \\
1.0000 \\
1.0000
\end{pmatrix}
\Rightarrow x =
\begin{pmatrix}
1.0000 \\
1.0000 \\
1.0000 \\
1.0000
\end{pmatrix} x_4
$$

while for $x_3 = -0.1691x_4$ one gets

$$\tilde{V}\begin{pmatrix} -0.1691x_4^2 \\ x_4^2 \end{pmatrix} = x_4^2 \begin{pmatrix} 1.6367 \\ 1.4780 \\ -0.2164 \\ 1.2793 \\ 1.3347 \\ -0.1954 \\ 1.1553 \\ 0.0286 \\ -0.1691 \\ 1.0000 \end{pmatrix} \Rightarrow x = \begin{pmatrix} 1.2793 \\ 1.1553 \\ -0.1691 \\ 1.0000 \end{pmatrix} x_4.$$

Then, it can be verified that the above solutions satisfy (1.75). Hence, one can conclude that

$$\mathcal{X} = \left\{ \frac{1}{2}\begin{pmatrix} 1.0000 \\ 1.0000 \\ 1.0000 \\ 1.0000 \end{pmatrix}, \frac{1}{2}\begin{pmatrix} 1.2793 \\ 1.1553 \\ -0.1691 \\ 1.0000 \end{pmatrix} \right\}.$$

1.9.2 Extended Procedure

When $u > m+1$, the procedure presented in the previous section cannot be applied, as it is not possible to express the coefficients s_i in (1.81) as homogeneous functions of only two variables. However, in the following it will be shown that if

$$u \leq (n-1)(m-1)+2 \tag{1.89}$$

this procedure can be suitably modified so that coefficients s_i can be written as rational functions of only one variable. This reduces once again the computation of $\mathcal{X}$ to the solution of a univariate polynomial equation.

To explain the basic idea, let us first consider the case $u = m+2$. Let us select three variables x_j, x_k and x_h with $1 \leq j \leq n$, $1 \leq k \leq n$, $1 \leq h \leq n$, $j \neq k$, $j \neq h$, $k \neq h$. Let us operate the pivot procedure in order to obtain a new matrix $\tilde{V}$ satisfying (1.82), where the rows $\tilde{r}_1, \ldots, \tilde{r}_u$ correspond to the monomials $x_k^m, x_j x_k^{m-1}, \ldots, x_j^m, x_h x_k^{m-1}$ in $x^{\{m\}}$. Then, it is straightforward to verify that in the new base

$$\tilde{s}_i = x_j^{m+2-i} x_k^{i-2}, \quad 2 \leq i \leq m+2$$
$$\tilde{s}_1 = x_h x_k^{m-1}.$$

Hence, any $v \in \text{img}(V)$ can be written as

$$v = x_h x_k^{m-1} \tilde{v}_1 + \sum_{i=2}^{m+2} x_j^{m+2-i} x_k^{i-2} \tilde{v}_i \tag{1.90}$$

which is as linear combination of the vectors $\tilde{v}_i$ weighted by x_h and powers of x_j, x_k.

Now, in order to eliminate x_h from (1.90), let us consider the a-th row of $\tilde{V}$ corresponding to monomial $x_j x_h x_k^{m-2}$. One has

$$x_j x_h x_k^{m-2} = \tilde{V}_{a,1} x_h x_k^{m-1} + \sum_{i=2}^{m+2} \tilde{V}_{a,i} x_j^{m+2-i} x_k^{i-2} \tag{1.91}$$

and hence, assuming $x_j \neq \tilde{V}_{a,1}$,

$$x_h x_k^{m-2} = \frac{\sum_{i=2}^{m+2} \tilde{V}_{a,i} x_j^{m+2-i} x_k^{i-2}}{x_j - \tilde{V}_{a,1} x_k} \tag{1.92}$$

which can be substituted into (1.90) (the case $x_j = \tilde{V}_{a,1}$ can be treated separately).

Therefore, each vector $v \in \mathcal{V}$ can be written as a rational function of the variables x_j and x_k only. Proceeding as in Section 1.9.1, one can exploit one relation between the monomials of $x^{\{m\}}$ given by some row of $\tilde{V}$ not used yet (e.g., the one corresponding to monomial $x_h^2 x_k^{m-2}$), to obtain a homogeneous equation in the variables x_j and x_k that all solutions of $\mathcal{X}$ must satisfy.

Let us turn now to the general case, to see how the above reasoning can be extended to all situations in which $m + 1 < u \leq (n-1)(m-1) + 2$. Without loss of generality, let us choose $j = 1$ and let i be the smallest integer less than n such that $u \leq m + 1 + i(m-1)$. Let us operate the pivot procedure in order to obtain a new matrix $\tilde{V}$ satisfying (1.82) where the rows $\tilde{r}_1, \ldots, \tilde{r}_u$ correspond to the monomials

$$x_k^m, x_1 x_k^{m-1}, \ldots, x_1^m,$$
$$x_2 x_k^{m-1}, x_1 x_2 x_k^{m-2}, \ldots, x_1^{m-2} x_2 x_k,$$
$$\vdots \tag{1.93}$$
$$x_i x_k^{m-1}, x_1 x_i x_k^{m-2}, \ldots, x_1^{m-2} x_i x_k,$$
$$x_{i+1} x_k^{m-1}, x_1 x_{i+1} x_k^{m-2}, \ldots, x_1^{l-1} x_{i+1} x_k^{m-i-l}$$

and l is such that $u = m + 1 + (i-1)(m-1) + l$. Then, in the new base the coefficients $\tilde{s}_i$ are polynomial functions of x_1 and linear in $x_2, x_3, \ldots, x_{i+1}$.

In order to eliminate all variables except x_1, thus obtaining $\tilde{s}$ as a function of x_1 only, let us consider the rows corresponding to the monomials

$$x_1^{m-1} x_2, \; x_1^{m-1} x_3, \; \ldots, \; x_1^l x_{i+1} x_k^{m-i-l-1}.$$

From these rows, following the same reasoning as in (1.91), it is easy to obtain the equation

$$A(x_1, x_k) z + B(x_1, x_k) = 0 \tag{1.94}$$

where $A(x_1,x_k) \in \mathbb{R}^{i \times i}$ and $B(x_1,x_k) \in \mathbb{R}^i$ are homogeneous functions of x_1 and x_k, and

$$z = (x_2, x_3, \ldots, x_{i+1})'. \tag{1.95}$$

From (1.94)–(1.95) it is possible to express $x_2, x_3, \ldots, x_{i+1}$ as rational functions of x_1 and x_k in the following way:

$$z = -A^{-1}(x_1, x_k)B(x_1, x_k) \tag{1.96}$$

(again, one can check separately the values of x_1 and x_k for which $A(x_1,x_k)$ is not invertible).

Then, each vector $v \in \text{img}(V)$ can be written as a rational homogeneous function of the variables x_1 and x_k only, by exploiting (1.95)–(1.96). Using one relation between the monomials of $x^{\{m\}}$, given by some row of $\tilde{V}$ not used yet, one can write a polynomial equation in the ratio $x_1 x_k^{m-1}$ that all solutions of $\mathscr{X}$ must satisfy.

Clearly, the above procedure can be applied only if the number of monomials in (1.93) is sufficient to cover all coefficients in $\tilde{s}$, or equivalently if u satisfies

$$u \le m + 1 + i(m-1)$$

for some $i < n - 1$. Therefore, it is easy to see that the maximum dimension of $\text{img}(V)$ for which the procedure can be applied is obtained with $i = n - 2$, which provides (1.89).

It is worth remarking that in the general case $u > (n-1)(m-1)+2$, the problem of finding a power vector within a given linear space requires the solution of a system of homogeneous polynomial equations. Nevertheless, such a case occurs rarely in practice, as we will see in the following chapters.

Example 1.15. Consider $n = 4$, $m = 2$ and

$$
x^{\{m\}} = \begin{pmatrix} x_1^2 \\ x_1 x_2 \\ x_1 x_3 \\ x_1 x_4 \\ x_2^2 \\ x_2 x_3 \\ x_2 x_4 \\ x_3^2 \\ x_3 x_4 \\ x_4^2 \end{pmatrix}, \quad V = \begin{pmatrix} 0.0320 & 0.2831 & 0.1991 & -0.2245 \\ 0.4811 & 0.5080 & 0.2440 & -0.0695 \\ -0.0796 & 0.2890 & 0.2520 & 0.1191 \\ -0.2840 & 0.3611 & 0.2328 & -0.0471 \\ -0.2216 & 0.2625 & -0.1472 & -0.7363 \\ -0.5497 & -0.3413 & 0.4525 & -0.1866 \\ 0.1828 & -0.1742 & 0.7382 & 0.0528 \\ 0.2855 & -0.4235 & -0.0196 & -0.4143 \\ 0.4623 & -0.2279 & 0.1059 & -0.2426 \\ 0.0240 & 0.0305 & 0.0081 & -0.3438 \end{pmatrix}.
$$

The problem is to compute the set $\mathscr{X}$ in (1.76).

Observe that $u = 4$ and hence the basic procedure cannot be applied being $u > m + 1$. However, $u \le (n-1)(m-1)+2 = 5$ and hence the extended procedure can be employed.

Let us select $j = 3$, $k = 4$ and $h = 2$. By performing the pivot procedure on matrix V with respect to monomials x_4, x_3x_4, x_3^2 and x_2x_4, one gets

$$
\tilde{V} = \begin{pmatrix}
0.1611 & -0.8934 & 0.4840 & 1.4128 \\
-0.1064 & -2.4058 & 2.4998 & 1.3209 \\
0.2788 & -0.8774 & 0.2288 & 0.5924 \\
0.3389 & -0.6887 & -0.3901 & 1.2943 \\
-0.1611 & -0.1066 & -0.4840 & 2.5872 \\
1.0621 & 1.9078 & -2.8073 & 0.3880 \\
1.0000 & 0.0000 & 0.0000 & 0.0000 \\
0.0000 & 1.0000 & 0.0000 & 0.0000 \\
0.0000 & 0.0000 & 1.0000 & 0.0000 \\
0.0000 & 0.0000 & 0.0000 & 1.0000
\end{pmatrix}
\begin{matrix}
\rightarrow x_1^2 \\
\rightarrow x_1x_2 \\
\rightarrow x_1x_3 \\
\rightarrow x_1x_4 \\
\rightarrow x_2^2 \\
\rightarrow x_2x_3 \\
\rightarrow x_2x_4 \\
\rightarrow x_3^2 \\
\rightarrow x_3x_4 \\
\rightarrow x_4^2
\end{matrix}
$$

and hence one has

$$
x^{\{m\}} = \tilde{V} \begin{pmatrix} x_2x_4 \\ x_3^2 \\ x_3x_4 \\ x_4^2 \end{pmatrix}. \tag{1.97}
$$

From the sixth row in (1.97), one gets

$$
x_2x_3 = 1.0621x_2x_4 + 1.9078x_3^2 - 2.8073x_3x_4 + 0.3380x_4^2
$$

and hence

$$
x_2 = \frac{1.9078x_3^2 - 2.8073x_3x_4 + 0.3380x_4}{x_3 - 1.0621x_4} \tag{1.98}
$$

which clearly corresponds to (1.92). Hence, each $v \in img(V)$ can be expressed as a rational function in x_3 and x_4. By considering the fifth row in (1.97), one has

$$
x_2^2 = -0.1611x_2x_4 - 0.1066x_3^2 - 0.4840x_3x_4 + 2.5872x_4^2.
$$

Substituting x_2 from (1.98), one finds

$$
\frac{3.7465x_3^4 - 10.1470x_3^3x_4 + 5.0879x_3^2x_4^2 + 4.4060x_3x_4^3 - 2.8344x_4^4}{(x_3 - 1.0621x_4)^2} = 0
$$

which gives the following solutions for x_3:

$$
x_3 \in \{1.5280x_4, 1.2107x_4, 0.6245x_4, -0.6548x_4\}.
$$

The corresponding solutions for x_2 are immediately obtained from (1.98). Then, in order to find x_1, one has just to compute the vector $x^{\{m\}}$ as in (1.97) for each of the four found pairs (x_2, x_3), and read the fourth entry corresponding to x_1x_4. Finally, one has just to verify that these solutions satisfy (1.75). This gives the following solution for $\mathscr{X}$:

$$
\mathscr{X} = \left\{ 0.4472 \begin{pmatrix} -0.5077 \\ 1.1864 \\ 1.5280 \\ 1.0000 \end{pmatrix}, \ 0.4472 \begin{pmatrix} -0.6760 \\ -1.4412 \\ 1.2107 \\ 1.0000 \end{pmatrix}, \right.
$$

$$
\left. 0.4472 \begin{pmatrix} 1.2630 \\ 1.4194 \\ 0.6245 \\ 1.0000 \end{pmatrix}, \ 0.4472 \begin{pmatrix} 0.6536 \\ -1.7732 \\ -0.6548 \\ 1.0000 \end{pmatrix} \right\}.
$$

1.10 Notes and References

The potential of forms in system and control theory has been recognized since long time, see e.g. [19, 13]. In recent years, this topic has gained a renewed interest, motivated by the strong connection with SDP and convex optimization techniques [15, 98].

The SMR of forms was introduced in [50, 51]. In the literature, it is also known as Gram matrix, see for instance [53] and references therein. Algorithms for the computation of the SMR have been provided in [36]. The SMR of matrix forms has been proposed in [35].

The use of SOS forms for studying positivity of forms has been widely investigated in the mathematical literature, see e.g. [121, 131, 129, 118, 53, 113, 132, 97]. The characterization of SOS forms via LMIs was proposed in [50, 51] and in [106, 107]. An alternative approach based on the theory of moments of a probability measure, also known in the literature as lifting, has been introduced in [85]. In terms of the resulting convex optimization problems, there is a duality relationship between the moments approach and the SOS-based approach. Conditions for a form to be SOS, not based on LMIs, have been given in [87].

Characterization of SOS matrix forms via LMIs have been proposed in [35] and in [77]. Alternative approaches for studying positivity of matrix forms have been proposed via the theory of moments [74] and via slack variables [110].

SOS-based relaxations are largely exploited in control systems, see for instance the tutorials [81, 105], the book [71], the special issue [45], and references therein. A large number of applications can be found in robust control [75, 78, 74, 127], analysis and design of nonlinear control systems [137, 130, 22, 21, 117, 39, 135, 138, 47], time-delay systems [104, 111], hybrid systems [115, 114], and many others. SOS forms have been employed also in other fields such as computer vision [44, 46, 28], robotics [30], and systems biology [61, 49].

The technique for investigating positivity of a polynomial over an ellipsoid, described in Section 1.7, was first proposed in [51] and then exploited in [52, 36]. The approach for establishing positivity of matrix polynomials on the simplex, described in Section 1.8, has been adopted in different contexts, see e.g. [22] and [42]. Different relaxations for the minimization of forms on the simplex have been investigated in [57].

The extraction of power vectors from linear spaces described in Section 1.9 has been proposed in [33] where it is shown how one can compute the solutions of systems of polynomial equations via LMIs. An alternative technique for extracting power vectors from linear spaces has been presented within the moments framework in [73], which relies on a result in [55].

There are several software tools which allow one to formulate and solve convex optimization problems relevant to positivity of forms and polynomials, see e.g. [134, 116, 72, 91].

Chapter 2
Positivity Gap

This chapter investigates the gap between positive forms and SOS forms. Conservatism of the LMI relaxations described in Chapter 1 is related to the existence of positive forms which are not SOS, called PNS forms. *a priori* conditions for non-conservatism of these relaxations are presented for some classes of forms. The class of SMR-tight forms is introduced in order to derive *a posteriori* tightness conditions. A further contribution of this chapter consists of providing a parametrization of the set of PNS forms. It is shown that the set of PNS forms is dense in the space of forms, that each PNS form is the vertex of a cone of PNS forms, and how PNS forms can be constructed via the SMR.

2.1 Hilbert's 17th Problem

Is it true that any positive semidefinite form is an SOS form? This question is closely related to Hilbert's 17th problem [122], which concerns the possibility of representing nonnegative polynomials as a sum of squares of rational functions.

The answer to the former question is negative. This fact was discovered by Hilbert himself in 1888 via a non-constructive proof [118]. In 1967, Motzkin provided an example of form which is positive semidefinite but not SOS. This form has degree 6 in 3 scalar variables, and is given by [122]

$$h_{Mot}(x) = x_1^4 x_2^2 + x_1^2 x_2^4 + x_3^6 - 3x_1^2 x_2^2 x_3^2. \tag{2.1}$$

Indeed, it can be verified that $h_{Mot}(x)$ is positive semidefinite and $h_{Mot}(x)$ is not SOS. In particular, one has that

$$\mu(h_{Mot}) = 0, \quad \lambda(h_{Mot}) = -0.0070.$$

Hence, there are forms that are positive semidefinite but not SOS. The following result, found by Artin in 1927, states that any positive semidefinite form is the ratio of two SOS forms [70].

G. Chesi et al.: Homogeneous Polynomial Forms, LNCIS 390, pp. 39–61.
springerlink.com © Springer-Verlag Berlin Heidelberg 2009

Theorem 2.1. *A form $h \in \Xi_{n,2m}$ is positive semidefinite if and only if there exist $h_1 \in \Sigma_{n,2(a+m)}$ and $h_2 \in \Sigma_{n,2a}$ for some integer $a \geq 0$, such that*

$$h(x) = \frac{h_1(x)}{h_2(x)}. \tag{2.2}$$

The following result, found by Polya in 1928, characterizes the forms that are positive on the simplex [70].

Theorem 2.2. *A form $h \in \Xi_{n,d}$ is positive on the simplex Y_n in (1.68) if and only if there exists an integer $k \geq 0$ such that the coefficients of*

$$h(x) \left(\sum_{i=1}^{n} x_i \right)^k \tag{2.3}$$

are positive.

In the sequel we will investigate forms that are positive semidefinite but not SOS. First of all, let us introduce the following definition.

Definition 2.1 (PNS). A form $h \in \Xi_{n,2m}$ is *PNS* if it is positive semidefinite but not SOS.

We will indicate the set of PNS forms of degree $2m$ in n scalar variables as follows:

$$\Delta_{n,2m} = \{ h \in \Xi_{n,2m} : h(x) \text{ is PNS} \}. \tag{2.4}$$

Therefore, the set $\Omega_{n,2m}$ in (1.51) can be expressed as

$$\Omega_{n,2m} = \Sigma_{n,2m} \cup \Delta_{n,2m}.$$

An interesting fact is that the set $\Delta_{n,2m}$ is empty for some values of n, m. Indeed, let us define the set

$$\mathscr{E} = \{ (n,2), \ n \in \mathbb{N} \} \cup \{ (2,2m), \ m \in \mathbb{N} \} \cup \{ (3,4) \}. \tag{2.5}$$

The following result states an important property of $\Delta_{n,2m}$ for any pair $(n,2m)$ in $\mathscr{E}$. A formal proof can be found in [70].

Theorem 2.3. *Let $(n,2m) \in \mathscr{E}$. Then, $\Delta_{n,2m} = \emptyset$, i.e. for all $h \in \Xi_{n,2m}$ one has*

$$h(x) \text{ is positive semidefinite } \iff \lambda(h) \geq 0. \tag{2.6}$$

The following result provides a further property of the forms in $\Xi_{n,2m}$ with $(n,2m)$ in $\mathscr{E}$, in particular stating that these forms are positive definite if and only if they admit a positive definite SMR matrix.

Theorem 2.4. *Let* $(n, 2m) \in \mathscr{E}$. *Then, for all* $h \in \Xi_{n,2m}$ *one has that*

$$h(x) \text{ is positive definite} \iff \lambda(h) > 0. \tag{2.7}$$

Proof. (Necessity) Let us suppose that $h(x)$ is positive definite. From Theorem 1.7 this means that $\mu(h) > 0$. Let us define the form

$$h_1(x) = h(x) - \mu(h)\|x^{\{m\}}\|^{2m}. \tag{2.8}$$

We have that

$$
\begin{aligned}
\mu(h_1) &= \min_{x \in \mathscr{C}_{n,m}} h_1(x) \\
&= \min_{x \in \mathscr{C}_{n,m}} \left(h(x) - \mu(h)\|x^{\{m\}}\|^{2m} \right) \\
&= \mu(h) - \mu(h) \\
&= 0.
\end{aligned}
$$

From Theorem 1.7 this implies that $h_1(x)$ is positive semidefinite. Moreover, $(n, 2m) \in \mathscr{E}$, and hence from (2.6) it follows that $h_1(x)$ is SOS. Therefore, we have that

$$0 \leq \lambda(h_1) \leq \mu(h_1) = 0$$

which implies that $\lambda(h_1) = 0$. From Theorem 1.4, $h_1(x)$ can be written as

$$h_1(x) = x^{\{m\}'} H_1 x^{\{m\}}$$

where $H_1 \in \mathbb{S}^{\sigma(n,m)}$ is positive semidefinite. Now, let us express $h(x)$ as $h(x) = x^{\{m\}'} H x^{\{m\}}$. It follows from (2.8) that

$$H = H_1 + \mu(h) I_{\sigma(n,m)}$$

which implies that

$$\lambda(h) = \lambda(h_1) + \mu(h) = \mu(h). \tag{2.9}$$

Since $\mu(h) > 0$, it follows that $\lambda(h) > 0$.

(Sufficiency) Let us suppose that $\lambda(h) > 0$. From Theorem 1.9 it follows that $\mu(h) \geq \lambda(h) > 0$. From Theorem 1.7 this implies that $h(x)$ is positive definite. $\square$

A direct consequence of Theorem 2.4 is that for forms with $(n, 2m) \in \mathscr{E}$, the SOS index coincides with the positivity index.

Corollary 2.1. *Let* $(n, 2m) \in \mathscr{E}$. *Then,*

$$\lambda(h) = \mu(h) \quad \forall h \in \Xi_{n,2m}. \tag{2.10}$$

Proof. It follows from (2.9) in the proof of Theorem 2.4. $\square$

Example 2.1. Let us consider the form

$$h(x) = x_1^4 + x_2^4 + x_3^4 - 3x_1^2 x_2 x_3. \tag{2.11}$$

One has that $h \in \Xi_{3,4}$. By solving the EVP (1.29), one gets $\lambda(h) = -0.0310$. By Theorem 1.4, this implies that $h(x)$ is not SOS. Moreover, since $(3,4) \in \mathscr{E}$, we can conclude from Theorem 2.3 that $h(x)$ is not positive semidefinite, i.e.

$$\exists x \in \mathbb{R}^3 : \ h(x) < 0.$$

Indeed, for $x_1 = 4/3$, $x_2 = 1$, $x_3 = 1$, one has $h(x) = -14/81$.

Example 2.2. Let us consider

$$h(x) = x_1^{2m} + x_2^{2m}, \quad m \in \mathbb{N}, \ m > 0. \tag{2.12}$$

We have that $h \in \Xi_{2,2m}$. Moreover, it is straightforward to verify that $h(x)$ is positive definite. Then, since $(2,2m) \in \mathscr{E}$ for any considered m, one can conclude from Theorem 2.4 that $\lambda(h) > 0$, or in other words, $h(x)$ admits a positive definite SMR matrix according to Lemma 1.2.

Throughout the book, the results in Theorems 2.3 and 2.4 will be exploited to formulate *a priori* conditions, which guarantee that results based on SOS relaxations of problems involving positivity of forms are not conservative.

As an example, let us consider the problem of checking positivity of a polynomial over an ellipsoid, addressed in Section 1.7. It can be observed that the LMI conditions in Theorem 1.16 are not only sufficient but also necessary, for some values of n, m. The next result is a direct consequence of Theorem 2.4.

Theorem 2.5. *Let $(n,2m) \in \mathscr{E}$. Then, the conditions in Theorem 1.16 are not only sufficient but also necessary for (1.59) to hold.*

Example 2.3. Let us consider the problem to establish whether (1.59) holds with

$$f(x) = 0.5 + x_1 + x_2^2, \quad Q = I_2, \quad c = 1.$$

The SOS index of the resulting $w(x;c)$ is negative, in particular $\lambda(w(\cdot;c)) = -0.7500$. Since $n = 2$ and $m = 2$ we have that $(n,2m) \in \mathscr{E}$. Therefore, from Theorems 1.16 and 2.5 we have that (1.59) does not hold, i.e. there exists some $x \in \mathscr{B}(Q,c)$ such that $f(x) \leq 0$.

2.2 Maximal SMR Matrices

In Chapter 1 it has been shown that a form may be represented by different SMR matrices. This section investigates the SMR matrices whose minimum eigenvalue coincides with the SOS index of the form. Roughly speaking, such matrices can be considered the "most positive definite" SMR matrices of the form. As it will be explained in the next section, these matrices are useful in order to study the gap between positive forms and SOS forms.

Definition 2.2 (Maximal SMR Matrix). Let $H^* \in \mathbb{S}^{\sigma(n,m)}$ be an SMR matrix of $h \in \Xi_{n,2m}$. Then, H^* is called a *maximal SMR matrix* of $h(x)$ if

$$\lambda_{min}(H^*) = \lambda(h). \tag{2.13}$$

Given a form $h(x)$, its maximal SMR matrices can be obtained as

$$H^* = H + L(\alpha^*) \tag{2.14}$$

where α^* is a value of α for which the maximum in (1.29) is achieved, and $H + L(\cdot)$ is the complete SMR matrix of $h(x)$ in (1.29).

2.2.1 Minimum Eigenvalue Decomposition

The following definition introduces a key decomposition of symmetric matrices which will be exploited in the sequel. For ease of presentation, the decomposition is formulated for a matrix of size $\sigma(n,m)$, though it can be defined for matrices of any size.

Definition 2.3 (Minimum Eigenvalue Decomposition). For a matrix $H \in \mathbb{S}^{\sigma(n,m)}$ we say that the quadruplet $\langle \lambda_{min}(H), \beta, V_0, V_p \rangle$ is a *minimum eigenvalue decomposition* of H if

$$H = VDV' \tag{2.15}$$

where $D \in \mathbb{S}^{\sigma(n,m)}$ is the diagonal matrix

$$D = \lambda_{min}(H)I_{\sigma(n,m)} + \text{diag}\begin{pmatrix} 0_{\sigma(n,m)-r} \\ \beta \end{pmatrix} \tag{2.16}$$

with

$$\begin{cases} \beta \in \mathbb{R}^r \\ \beta > 0 \end{cases} \tag{2.17}$$

and $V \in \mathbb{R}^{\sigma(n,m) \times \sigma(n,m)}$ is an orthogonal matrix such that

$$\begin{cases} V = \begin{pmatrix} V_0 & V_p \end{pmatrix} \\ V_0 \in \mathbb{R}^{\sigma(n,m) \times (\sigma(n,m)-r)}, \quad V_p \in \mathbb{R}^{\sigma(n,m) \times r} \\ VV' = V'V = I_{\sigma(n,m)}. \end{cases} \tag{2.18}$$

It follows that the diagonal of D contains the eigenvalues of H, V is a matrix of eigenvectors, and r is an integer satisfying $1 \leq r \leq \sigma(n,m)$ which represents the number of eigenvalues of H distinct from $\lambda_{min}(H)$ (including their multiplicity).

It is useful to observe that $\langle \lambda_{min}(H), \beta, V_0, V_p \rangle$ is a minimum eigenvalue decomposition of H if and only if $\langle \lambda_{min}(H), T_1\beta, V_0T_2, V_pT_1^{-1} \rangle$ is, for all matrices $T_1 \in \mathbb{R}^{r \times r}$ and $T_2 \in \mathbb{R}^{(\sigma(n,m)-r) \times (\sigma(n,m)-r)}$ such that T_1 is a permutation matrix and T_2 is a nonsingular matrix.

2.2.2   Structure of Maximal SMR Matrices

The following result provides a fundamental property of maximal SMR matrices.

Theorem 2.6. *Let* $h \in \Xi_{n,2m}$, $H + L(\alpha)$ *be a complete SMR matrix of* $h(x)$, *and* $\langle \lambda_{min}(H), \beta, V_0, V_p \rangle$ *be a minimum eigenvalue decomposition of* H. *Let us define*

$$\eta^*(V_0) = \max_{\alpha: \|\alpha\|=1} \lambda_{min}\left(V_0'L(\alpha)V_0\right). \tag{2.19}$$

Then, H is a maximal SMR matrix of h if and only if

$$\eta^*(V_0) \leq 0. \tag{2.20}$$

Proof. From (2.13) it follows that H is a maximal SMR matrix if and only if

$$\lambda_{min}(H + L(\alpha)) \leq \lambda_{min}(H) \quad \forall \alpha$$

and, hence, if and only if

$$\forall \alpha \; \exists y, \|y\| = 1 : \; y'(H + L(\alpha))y \leq \lambda_{min}(H). \tag{2.21}$$

Let $\langle \lambda_{min}(H), \beta, V_0, V_p \rangle$ be a minimum eigenvalue decomposition of H. Then, (2.21) can be rewritten as

$$\forall \alpha \; \exists y, \|y\| = 1 : \; y'V_p \text{diag}(\beta)V_p'y \leq -y'L(\alpha)y. \tag{2.22}$$

Let us observe that $L(\alpha)$ depends linearly on α. This means that $V_p'y$ tends to zero as α tends to zero because

$$\text{diag}(\beta) > 0.$$

Moreover, if (2.22) holds for the pair (y, α), it also holds for the pair $(y, c\alpha)$ for all $c \geq 1$. Therefore, it turns out that H is a maximal SMR matrix if and only if

$$\forall \alpha \; \forall \varepsilon > 0 \; \exists y, \|y\| = 1 : \; \|V_p'y\| < \varepsilon \text{ and } y'V_p \text{diag}(\beta)V_p'y \leq -y'L(\alpha)y$$

or, equivalently, if and only if

$$\forall \alpha \, \exists y, \|y\| = 1 : \, V_p' y = 0_r \text{ and } y' V_p \, \text{diag}(\beta) V_p' y \leq -y' L(\alpha) y. \tag{2.23}$$

Let us observe that

$$\ker(V_p') = \text{img}(V_0) \tag{2.24}$$

and hence

$$V_p' y = 0 \iff y \in \text{img}(V_0).$$

Therefore, (2.23) can be rewritten as

$$\forall \alpha \, \exists y \in \text{img}(V_0), \|y\| = 1 : \, y' L(\alpha) y \leq 0. \tag{2.25}$$

Let us observe that

$$y \in \text{img}(V_0) \iff y = V_0 p, \quad p \in \mathbb{R}^{\sigma(n,m)-r}.$$

Since $y' L(\alpha) y$ depends linearly on α, the condition (2.25) can be rewritten as

$$\forall \alpha, \|\alpha\| = 1, \, \exists p, \|p\| = 1 : \, p' V_0' L(\alpha) V_0 p \leq 0$$

which is equivalent to (2.20). $\qquad\qquad\qquad\qquad\qquad\qquad\qquad\qquad\qquad\square$

Theorem 2.6 provides a necessary and sufficient condition to establish if a given SMR matrix H is a maximal SMR matrix. This condition is important because it states that the property of being a maximal SMR matrix is related only to the matrix V_0 in the minimum eigenvalue decomposition of H, which represents the eigenspace of the minimum eigenvalue of H. In particular, this eigenspace is given by $\text{img}(V_0)$. Hence, Theorem 2.6 provides a way to construct maximal SMR matrices.

Let us observe that the feasible set for α in (2.19) is nonconvex, which makes the computation of the index $\eta^*(V_0)$ difficult. The following result provides an alternative way for characterizing maximal SMR matrices.

Theorem 2.7. *Let V_0 and $L(\alpha)$ be defined as in Theorem 2.6, and define*

$$\eta(V_0) = \max \, \{\eta(V_0, 1), \eta(V_0, -1)\} \tag{2.26}$$

where

$$\eta(V_0, z) = \sup_{\alpha: \, y'\alpha = z} \lambda_{min} \left(V_0' L(\alpha) V_0 \right) \tag{2.27}$$

and $y \in \mathbb{R}_0^{\omega(n,m)}$. Then, for all $y \in \mathbb{R}_0^{\omega(n,m)}$, one has

$$\eta^*(V_0) \leq 0 \iff \eta(V_0) \leq 0. \tag{2.28}$$

Proof. (Necessity) Let us assume that $\eta^*(V_0) \leq 0$ and let us suppose by contradiction that $\eta(V_0) > 0$. Then, there exists $\tilde{\alpha} \in \mathbb{R}^{\omega(n,m)}$ such that $|y'\tilde{\alpha}| = 1$ and

$$\lambda_{min}\left(V_0'L(\tilde{\alpha})V_0\right) > 0. \tag{2.29}$$

Let us define

$$\bar{\alpha} = \|\tilde{\alpha}\|^{-1}\tilde{\alpha}.$$

We have that $\|\bar{\alpha}\| = 1$ and

$$\lambda_{min}\left(V_0'L(\bar{\alpha})V_0\right) = \|\tilde{\alpha}\|^{-1}\lambda_{min}\left(V_0'L(\tilde{\alpha})V_0\right) > 0.$$

But this is impossible since we have assumed that $\eta^*(V_0) \le 0$.

(Sufficiency) Let us assume that $\eta(V_0) \le 0$ and let us suppose by contradiction that $\eta^*(V_0) > 0$. Then, there exists $\tilde{\alpha} \in \mathbb{R}^{\omega(n,m)}$ such that $\|\tilde{\alpha}\| = 1$ and (2.29) holds. First, let us suppose that

$$y'\tilde{\alpha} \ne 0 \tag{2.30}$$

and let us define

$$\bar{\alpha} = |y'\tilde{\alpha}|_2^{-1}\tilde{\alpha}.$$

We have that $\|y'\bar{\alpha}\| = 1$ and

$$\lambda_{min}\left(V_0'L(\bar{\alpha})V_0\right) = |w'\tilde{\alpha}|_2^{-1}\lambda_{min}\left(V_0'L(\tilde{\alpha})V_0\right) > 0.$$

But this is impossible since we have assumed that $\eta(V_0) \le 0$.
Now, let us suppose that

$$y'\tilde{\alpha} = 0.$$

Then, for all $\varepsilon > 0$ there exists $\hat{\alpha} \in \mathbb{R}^{\omega(n,m)}$ such that $\|\hat{\alpha}\| = 1$ and

$$\|\hat{\alpha} - \tilde{\alpha}\| < \varepsilon \text{ and } y'\hat{\alpha} \ne 0.$$

Since the function $\lambda_{min}\left(V_0'L(\alpha)V_0\right)$ is continuous with respect to α and since $\hat{\alpha}$ is arbitrarily close to $\tilde{\alpha}$ which satisfies (2.29), it follows that $\hat{\alpha}$ can be chosen to satisfy also the condition $\lambda_{min}\left(V_0'L(\hat{\alpha})V_0\right) > 0$. By repeating the proof from (2.30) by using $\hat{\alpha}$ instead of $\tilde{\alpha}$, we finally conclude that (2.28) holds. $\qquad\square$

Theorem 2.7 provides an alternative way to establish whether an SMR matrix is a maximal SMR matrix or not. This is achieved via the index $\eta(V_0)$, which can be computed through two convex optimizations. In fact, it turns out that $\eta(V_0, z)$ is the solution of the EVP

$$\eta(V_0, z) = \sup_{t,\alpha} t$$
$$\text{s.t.} \begin{cases} y'\alpha - z = 0 \\ V_0'L(\alpha)V_0 - tI_{\sigma(n,m)-r} \ge 0. \end{cases} \tag{2.31}$$

Let us observe that the free vector y defines two hyperplanes on which the function $\lambda_{min}\left(V_0'L(\alpha)V_0\right)$ is evaluated.

Example 2.4. Let us consider the form $h(x)$ in (1.20) and its SMR in (1.21). It can be verified that a minimum eigenvalue decomposition $\langle \lambda_{min}(H), \beta, V_0, V_p \rangle$ of the SMR matrix H in (1.21) is given by

$$\lambda_{min}(H) = -0.6180$$
$$\beta = (2.2361, 2.6180)'$$
$$V_0 = (0.5257, -0.8507, 0)'$$
$$V_p = \begin{pmatrix} -0.8507 & -0.5257 & 0 \\ 0 & 0 & 1.0000 \end{pmatrix}'.$$

By applying (2.26)-(2.27), we find that $\eta(V_0) > 0$, which implies from Theorem 2.7 that H is not a maximal SMR matrix. This is confirmed by the fact that there exists another SMR matrix of $h(x)$ whose minimum eigenvalue is larger than the minimum eigenvalue of H. This SMR matrix is given by H^* in (2.14) with $\alpha^* = 0.8008$, which is an optimal value of α in the EVP (1.29). Indeed we have:

$$H^* = \begin{pmatrix} 1.0000 & 1.0000 & -0.8008 \\ \star & 1.6016 & 0.0000 \\ \star & \star & 2.0000 \end{pmatrix} \tag{2.32}$$

and

$$\lambda_{min}(H^*) = 0.0352, \quad \lambda_{min}(H) = -0.6180.$$

Lastly, we test Theorem 2.7 on the SMR matrix H^*. To this end, consider the minimum eigenvalue decomposition of H^* given by

$$\lambda_{min}(H^*) = 0.0352$$
$$\beta^* = (1.7895, 2.7065)'$$
$$V_0^* = (0.7972, -0.5089, 0.3249)' \tag{2.33}$$
$$V_p^* = \begin{pmatrix} 0.1544 & 0.6920 & 0.7052 \\ -0.5837 & -0.5120 & 0.6302 \end{pmatrix}'.$$

From (2.26) we find $\eta(V_0^*) = 0.0000$ by solving (2.31) with $y = 1$, which verifies by Theorems 2.6 and 2.7 that H^* is a maximal SMR matrix.

2.3 SMR-tight Forms

This section introduces and characterizes a special class of forms, specifically the forms whose positivity index coincides with their SOS index.

Definition 2.4 (SMR-tight Form). Let us suppose $h \in \Xi_{n,2m}$ satisfies

$$\lambda(h) = \mu(h). \tag{2.34}$$

Then, $h(x)$ is said to be *SMR-tight*.

Before proceeding with the characterization of SMR-tight forms, let us make the following observations:

1. PNS forms are not SMR-tight. In fact, if $h(x)$ is PNS then $\mu(h) \geq 0$ and $\lambda(h) < 0$.
2. A form can be SMR-tight even if it is not SOS. This is shown by the following example.

Example 2.5. Let us consider the form

$$h(x) = x_1^2 + 4x_1x_2 + x_2^2.$$

We have that a complete SMR of $h(x)$ is given by

$$x^{\{m\}} = \begin{pmatrix} x_1 & x_2 \end{pmatrix}', \quad H = \begin{pmatrix} 1 & 2 \\ \star & 1 \end{pmatrix}, \quad L(\alpha) = 0_{2 \times 2}$$

which implies that the SOS index of $h(x)$ is

$$\lambda(h) = \lambda_{min}(H) = -1.$$

Then, it can be verified that the positivity index of $h(x)$ is

$$\begin{aligned} \mu(h) &= \min_{x \in \mathscr{C}_{n,m}} h(x) \\ &= \min_{x:\, x_1^2 + x_2^2 = 1} h(x) \\ &= -1. \end{aligned}$$

Therefore, $h(x)$ is SMR-tight because $\lambda(h) = \mu(h)$. However, $h(x)$ is not SOS: indeed, $\mu(h)$ is negative, which means that $h(x)$ can take negative values.

2.3.1 Minimal Point Set

A necessary and sufficient condition for establishing whether a form is SMR-tight can be obtained by searching for power vectors in a linear space. To this end, let us introduce the following definition.

Definition 2.5 (Minimal Point Set). Let $h \in \Xi_{n,2m}$, $H \in \mathbb{S}^{\sigma(n,m)}$ be a maximal SMR matrix of $h(x)$, and define the linear space

$$\mathscr{N}(H) = \ker\left(H - \lambda_{min}(H)I_{\sigma(n,m)}\right). \tag{2.35}$$

Then, the set

$$\text{mps}(h) = \left\{ x \in \mathbb{R}^n : \|x\| = 1,\ x^{\{m\}} \in \mathcal{N}(H) \right\} \tag{2.36}$$

is called *minimal point set* of $h(x)$.

The following lemma clarifies the relationship between $\mathcal{N}(H)$ and the minimal eigenvalue decompositions of H.

Lemma 2.1. *Let* $h \in \Xi_{n,2m}$, *and* $H \in \mathbb{S}^{\sigma(n,m)}$ *be a maximal SMR matrix of* $h(x)$. *Let* $\langle \lambda_{min}(H), \beta, V_0, V_p \rangle$ *be a minimum eigenvalue decomposition of* H. *Then,*

$$\mathcal{N}(H) = \text{img}(V_0). \tag{2.37}$$

Proof. From Definition 2.3 we have that the columns of V_0 are a base of the eigenspace of the minimum eigenvalue of H, which is $\mathcal{N}(H)$ according to (2.35). Therefore, (2.37) holds. □

It is worthwhile to observe that the minimal point set of $h(x)$ does not depend on the chosen maximal SMR matrix H. This is explained in the following result.

Theorem 2.8. *Let* $h \in \Xi_{n,2m}$, *and for* $i = 1,2$ *define*

$$\mathscr{A}_i = \left\{ x \in \mathbb{R}^n : \|x\| = 1,\ x^{\{m\}} \in \mathcal{N}(H_i) \right\}$$

where $H_1, H_2 \in \mathbb{S}^{\sigma(n,m)}$ *are any pair of maximal SMR matrices of* $h(x)$. *Then,*

$$\mathscr{A}_1 = \mathscr{A}_2$$

i.e. mps(h) *is independent on the chosen maximal SMR matrix* H *of* $h(x)$.

Proof. Let us suppose by contradiction that there exists $\bar{x} \in \mathscr{A}_1$ such that $\bar{x} \notin \mathscr{A}_2$. Since $\bar{x} \in \mathscr{A}_1$ we have that

$$\begin{aligned}
0 &= \bar{x}^{\{m\}'} \left(H_1 - \lambda_{min}(H_1) I_{\sigma(n,m)} \right) \bar{x}^{\{m\}} \\
&= h(\bar{x}) - \lambda_{min}(H_1) \left\| \bar{x}^{\{m\}} \right\|^2 .
\end{aligned} \tag{2.38}$$

Since H_2 is a maximal SMR matrix of $h(x)$ we have that

$$\begin{aligned}
h(\bar{x}) &= \bar{x}^{\{m\}'} H_2 \bar{x}^{\{m\}} \\
\lambda_{min}(H_1) &= \lambda_{min}(H_2)
\end{aligned}$$

which, from (2.38), provides

$$0 = \bar{x}^{\{m\}'} \left(H_2 - \lambda_{min}(H_2) I_{\sigma(n,m)} \right) \bar{x}^{\{m\}} .$$

Moreover,

$$H_2 - \lambda_{min}(H_2)I_{\sigma(n,m)} \geq 0$$

which implies that

$$\|\bar{x}\| = 1 \text{ and } \bar{x}^{\{m\}} \in \mathcal{N}(H_2)$$

hence contradicting the assumption $\bar{x} \notin \mathscr{A}_2$. $\qquad\qquad\square$

The following result states that a necessary and sufficient condition for a form to be SMR-tight is that the minimal point set of the form is not empty.

Theorem 2.9. *Let* $h \in \Xi_{n,2m}$. *Then,* $h(x)$ *is SMR-tight if and only if*

$$\mathrm{mps}(h) \neq \emptyset. \qquad (2.39)$$

Proof. (Sufficiency) Let us suppose that $\mathrm{mps}(h) \neq \emptyset$. Let $\bar{x}$ be any vector in $\mathrm{mps}(h)$ and define

$$\hat{x} = \frac{\bar{x}}{\|\bar{x}^{\{m\}}\|}.$$

By letting $H \in \mathbb{S}^{\sigma(n,m)}$ be a maximal SMR matrix of $h(x)$, we have that

$$\begin{aligned}
0 &= \hat{x}^{\{m\}'} \left(H - \lambda_{min}(H)I_{\sigma(n,m)} \right) \hat{x}^{\{m\}} \\
&= h(\hat{x}) - \lambda_{min}(H) \|\hat{x}^{\{m\}}\|^2 \\
&= h(\hat{x}) - \lambda(h)
\end{aligned}$$

which implies that

$$\exists \hat{x} \in \mathscr{C}_{n,m} : h(\hat{x}) = \lambda(h). \qquad (2.40)$$

By Definition 1.14, $\mu(h)$ is the minimum of $h(x)$ over the set $\mathscr{C}_{n,m}$; moreover, by Theorem 1.9, $\lambda(h)$ is a lower bound of $\mu(h)$. Therefore, from (2.40) we conclude that $\mu(h) = \lambda(h)$, i.e. $h(x)$ is SMR-tight.

(Necessity) Let us suppose that $h(x)$ is SMR-tight, i.e. $\mu(h) = \lambda(h)$. Then, (2.40) is satisfied. Let $H \in \mathbb{S}^{\sigma(n,m)}$ be a maximal SMR matrix of $h(x)$. We have that:

$$\begin{aligned}
0 &= h(\hat{x}) - \lambda(h) \\
&= h(\hat{x}) - \lambda_{min}(H) \|\hat{x}^{\{m\}}\|^2 \\
&= \hat{x}^{\{m\}'} \left(H - \lambda_{min}(H)I_{\sigma(n,m)} \right) \hat{x}^{\{m\}}.
\end{aligned}$$

Since $H - \lambda_{min}(H)I_{\sigma(n,m)} \geq 0$ it follows that there exists $\hat{x} \in \mathbb{R}_0^n$ such that $x^{\{m\}}$ belongs to $\mathcal{N}(H)$. Therefore, let us define

$$\bar{x} = \frac{\hat{x}}{\|\hat{x}\|}.$$

We have that $\bar{x} \in \mathrm{mps}(h)$, and hence (2.39) holds. $\qquad\qquad\square$

2.3.2 Rank Conditions

Clearly, it is possible to establish whether mps(h) is empty or not by computing vectors in mps(h) through the extraction procedure described in Section 1.9. In the sequel, we aim to provide alternative conditions for establishing whether mps(h) is empty, which do not require the actual computation of the set mps(h) itself.

Theorem 2.10. *Let $h \in \Xi_{n,2m}$, $H \in \mathbb{S}^{\sigma(n,m)}$ be a maximal SMR matrix of $h(x)$, and $\mathcal{N}(H)$ be the linear space in (2.35). Let us suppose that one of the following conditions holds:*

1. m is odd and $\dim(\mathcal{N}(H)) > \sigma(n,m) - n$;
2. m is even and $\dim(\mathcal{N}(H)) = \sigma(n,m)$.

Then, $h(x)$ is SMR-tight.

Proof. Let us suppose that item 1 holds, and let $\langle \lambda_{min}(H), \beta, V_0, V_p \rangle$ be a minimum eigenvalue decomposition of H. Let us consider the equation

$$V_p' x^{\{m\}} = 0. \tag{2.41}$$

We have that (2.41) defines a system of $\sigma(n,m) - \dim(\mathcal{N}(H))$ homogeneous equations of degree m in n scalar variables. In particular, the degree of these homogeneous equations is odd, and their number is smaller than the number of scalar variables because

$$\sigma(n,m) - \dim(\mathcal{N}(H)) < n.$$

This implies that

$$\exists x \in \mathbb{R}^n : \|x\| = 1, \ V_p' x^{\{m\}} = 0.$$

From Definition 2.3, $\text{img}(V_0) = \ker(V_p')$, and hence it immediately follows that

$$\exists x \in \mathbb{R}^n : \|x\| = 1, \ x^{\{m\}} \in \text{img}(V_0).$$

Moreover, by Lemma 2.1, $\text{img}(V_0) = \mathcal{N}(H)$, which implies that there exists $x \in \mathbb{R}^n$ such that $\|x\| = 1$ and $x^{\{m\}} \in \mathcal{N}(H)$. This means that mps($h$) $\neq \emptyset$ from Definition 2.5, and hence $h(x)$ is SMR-tight by Theorem 2.9.

Lastly, let us suppose that item 2 holds. It immediately follows that

$$\mathcal{N}(H) = \mathbb{R}^{\sigma(n,m)}$$

and hence

$$\text{mps}(h) = \{x \in \mathbb{R}^n : \|x\| = 1\}$$

i.e. mps(h) $\neq \emptyset$ and hence $h(x)$ is SMR-tight by Definition 2.5 and Theorem 2.9. $\square$

Theorem 2.10 provides a simple condition to establish whether a form is SMR-tight, which consists only in checking whether the dimension of the linear space $\mathcal{N}(H)$ lies in a given range.

Example 2.6. Let us consider the form $h(x)$ in Example 1.2. A maximal SMR matrix of $h(x)$ has been found in Example 2.4 and is given by H^* in (2.32). From Lemma 2.1 and (2.33) we have that the linear space $\mathcal{N}(H^*)$ is given by

$$\mathcal{N}(H^*) = \text{img} \begin{pmatrix} 0.7972 \\ -0.5089 \\ 0.3249 \end{pmatrix}.$$

Since H^* is constructed with respect to the power vector $x^{\{m\}} = (x_1^2, x_1 x_2, x_2^2)'$, it can be verified from Definition 2.5 that

$$\text{mps}(h) = \left\{ \pm \begin{pmatrix} 0.8429 \\ -0.5381 \end{pmatrix} \right\}.$$

Therefore, mps(h) is not empty, and hence $h(x)$ is SMR-tight according to Theorem 2.9.

Example 2.7. Let us consider the form

$$
\begin{aligned}
h(x) = {} & 27x_1^4 x_2^2 - 36\sqrt{3}x_1^2 x_2^4 + 72x_1^2 x_2^2 x_3^2 + 36x_2^6 - 24\sqrt{3}x_2^4 x_3^2 + 12x_2^2 x_3^4 \\
& + 12x_1^4 x_3^2 - 24\sqrt{3}x_1^2 x_3^4 + 27x_2^4 x_3^2 - 36\sqrt{3}x_2^2 x_3^4 + 36x_3^6.
\end{aligned}
$$

We have $n = 3$, $m = 3$ and $\sigma(n,m) = 10$. After computing a maximal SMR matrix H of $h(x)$, we find that $\dim(\mathcal{N}(H)) = 8$. Let us observe that mps(h) cannot be computed via the extraction procedure described in Section 1.9 because (1.89) does not hold, indeed $8 = u \not\leq (n-1)(m-1) + 2 = 6$. Then, let us consider Theorem 2.10. We have that m is odd, and the first condition of the theorem holds since $8 = \dim(\mathcal{N}(H)) > \sigma(n,m) - n = 7$. This implies that $h(x)$ is SMR-tight and hence the positivity index $\mu(h)$ is equal to the SOS index $\lambda(h)$, which in this case is equal to 0.

Example 2.8. Let us consider Motzkin's form in (2.1). We have $n = 3$, $m = 3$ and $\sigma(n,m) = 10$. After computing a maximal SMR matrix H of $h_{Mot}(x)$, we find that $\dim(\mathcal{N}(H)) = 7$. Let us observe that mps(h) cannot be computed via the extraction procedure described in Section 1.9 because (1.89) does not hold, indeed $7 = u \not\leq (n-1)(m-1) + 2 = 6$. Then, let us consider Theorem 2.10. We have that m is odd, however the first condition of the theorem does not hold since $7 = \dim(\mathcal{N}(H)) \not> \sigma(n,m) - n = 7$. This means that we cannot conclude that $h_{Mot}(x)$ is SMR-tight. This is in accordance with the fact that $h_{Mot}(x)$ cannot be SMR-tight since it is PNS, which implies $\lambda(h_{Mot}) < 0$ and $\mu(h_{Mot}) = 0$.

2.4 Characterizing PNS Forms via the SMR

This section investigates the structure of PNS forms through the SMR. In particular, it is shown that each PNS form is the vertex of a cone of PNS forms. Moreover, a parametrization of PNS forms is proposed.

2.4.1 Basic Properties of PNS Forms

First of all, let us observe that, while the sets $\Omega_{n,2m}$ and $\Sigma_{n,2m}$ are convex, the set $\Delta_{n,2m}$ is nonconvex. This is shown by the following example.

Example 2.9. Let us consider Motzkin's form in (2.1) and Stengle's form [122]

$$h_{Ste}(x) = x_1^3 x_3^3 + (x_2^2 x_3 - x_1^3 - x_1 x_3^2)^2, \qquad (2.42)$$

which are both in $\Delta_{3,6}$. Let us define the form

$$h(x) = \frac{1}{2}\left(h_{Mot}(x) + h_{Ste}(x)\right).$$

It can be verified that

$$\lambda(h) = 0$$

which means that $h(x)$ is SOS. Therefore, $h \notin \Delta_{3,6}$, which implies that $\Delta_{3,6}$ is not convex.

For any $h \in \Xi_{n,m}$ let us define the ball in $\Xi_{n,m}$ with radius $\delta \in \mathbb{R}$ centered in $h(x)$ as

$$\mathscr{B}_\delta(h) = \{h_1 \in \Xi_{n,m} : d(h_1, h) \leq \delta\} \qquad (2.43)$$

where $d : \Xi_{n,m} \times \Xi_{n,m} \to \mathbb{R}$ is the distance in $\Xi_{n,m}$ defined as

$$d(h_1, h) = \|g_1 - g\| \qquad (2.44)$$

being $g_1, g \in \mathbb{R}^{\sigma(n,m)}$ vectors representing respectively h_1, h according to the power vector representation (1.6).

The following result introduces some key properties of $\Delta_{n,2m}$.

Theorem 2.11. *Suppose that $\Delta_{n,2m} \neq \emptyset$. Then:*

1. there exists $h \in \Delta_{n,2m}$ such that $\mu(h) > 0$;
2. for any $h \in \Delta_{n,2m}$ such that $\mu(h) > 0$, it follows that

$$\exists \delta > 0 : \ \mathscr{B}_\delta(h) \subset \Delta_{n,2m}; \qquad (2.45)$$

3. for any $h \in \Delta_{n,2m}$ there exists $\delta > 0$ such that

$$\mathscr{B}_\delta(h) \cap \Omega_{n,2m} \subset \Delta_{n,2m}. \qquad (2.46)$$

Proof. Consider item 1, and let $\Delta_{n,2m} \neq \emptyset$. Then, there exists $h \in \Delta_{n,2m}$ such that $\mu(h) \geq 0$. Let us suppose that $\mu(h) = 0$ and let us define

$$h_1(x) = h(x) + \varepsilon x^{\{m\}'} x^{\{m\}}.$$

It follows that

$$\mu(h_1) = \mu(h) + \varepsilon = \varepsilon.$$

Moreover, let H be an SMR matrix of $h(x)$. We have that

$$H_1 = H + \varepsilon I_{\sigma(n,m)}$$

is an SMR matrix of $h_1(x)$. Hence, it follows that

$$\lambda(h_1) = \lambda(h) + \varepsilon.$$

Since $\lambda(h) < 0$, by choosing $\varepsilon \in (0, -\lambda(h))$, one gets $h_1 \in \Delta_{n,2m}$ and $\mu(h_1) > 0$. Hence, item 1 holds.

Consider item 2, and let $h \in \Delta_{n,2m}$ with $\mu(h) > 0$. We have also $\lambda(h) < 0$. For continuity of $\mu(h)$ and $\lambda(h)$ with respect to the coefficients of $h(x)$, it follows that

$$\exists \delta > 0: \ \mu(h_1) > 0 \text{ and } \lambda(h_1) < 0 \ \ \forall h_1 \in \mathscr{B}_\delta(h)$$

i.e. (2.45) holds.

Lastly, consider item 3, and let $h \in \Delta_{n,2m}$. If $\mu(h) > 0$, then (2.45) holds, which directly implies (2.46) since $\Delta_{n,2m} \subset \Omega_{n,2m}$. Hence, let us suppose $\mu(h) = 0$. Similarly to the proof of (2.45) it follows that

$$\exists \delta > 0: \ \lambda(h_1) < 0 \ \ \forall h_1 \in \mathscr{B}_\delta(h)$$

i.e. $\mathscr{B}_\delta(h) \cap \Sigma_{n,2m} = \emptyset$. Therefore, (2.46) holds. $\square$

Theorem 2.11 states three properties for the set of PNS forms $\Delta_{n,2m}$. The first says that, if this set is not empty, then it contains positive definite forms. The second property says that positive definite forms in $\Delta_{n,2m}$ are interior points of $\Delta_{n,2m}$. The third property establishes that every PNS form owns a neighborhood with shape defined by (2.43)–(2.44) where all positive semidefinite forms are PNS. This means that arbitrarily small changes of the coefficients of a PNS form cannot turn this form into an SOS form.

As it has been explained in the previous sections, to establish whether a form $h(x)$ is PNS amounts to establishing whether $\mu(h) \geq 0$ and $\lambda(h) < 0$. The following result provides a further characterization of PNS forms which turns out to be useful for their construction.

Theorem 2.12. *Let $h \in \Delta_{n,2m}$, and $H \in \mathbb{S}^{\sigma(n,m)}$ be a maximal SMR matrix of $h(x)$. Let $\langle \lambda_{min}(H), \beta, V_0, V_p \rangle$ be a minimum eigenvalue decomposition of H. Then,*

$$\nexists x \in \mathbb{R}_0^n: \ V_p' x^{\{m\}} = 0. \tag{2.47}$$

Proof. Let us suppose by contradiction that there exists $\tilde{x} \in \mathbb{R}_0^n$ such that $\tilde{x}^{\{m\}} \in \ker(V_p')$. Let r be the length of β. Then, we have

$$h(\tilde{x}) = \tilde{x}^{\{m\}'} \begin{pmatrix} V_0 & V_p \end{pmatrix} \left(\lambda_{min}(H) I_{\sigma(n,m)} + \operatorname{diag} \begin{pmatrix} 0_{\sigma(n,m)-r} \\ \beta \end{pmatrix} \right) \begin{pmatrix} V_0' \\ V_p' \end{pmatrix} \tilde{x}^{\{m\}}$$

$$= \lambda_{min}(H) \|V_0' \tilde{x}^{\{m\}}\|^2.$$

Let us observe that $\lambda_{min}(H) < 0$, because H is a maximal SMR matrix of a PNS form. Moreover, $\|V_0' \tilde{x}^{\{m\}}\| > 0$, since $\operatorname{img}(V_0) = \ker(V_p')$. This implies that $h(\tilde{x}) < 0$, which is impossible since $h(x)$ is PNS. $\square$

Theorem 2.12 provides a necessary condition for a form to be PNS: the absence of solutions $x \in \mathbb{R}_0^n$ in the homogeneous polynomial system $V_p' x^{\{m\}} = 0$. By Definition 2.3, this condition is equivalent to

$$\nexists x \in \mathbb{R}_0^n : x^{\{m\}} \in \operatorname{img}(V_0).$$

2.4.2 Cones of PNS Forms

The following result provides a way to generate a set of PNS forms from a given PNS form.

Theorem 2.13. *Let $h \in \Delta_{n,2m}$, $H \in \mathbb{S}^{\sigma(n,m)}$ be a maximal SMR matrix of $h(x)$, and $\langle \lambda_{min}(H), \beta, V_0, V_p \rangle$ be a minimum eigenvalue decomposition of H. Let us define the parametrized form*

$$s(x; \gamma) = x^{\{m\}'} V_p \operatorname{diag}(\gamma) V_p' x^{\{m\}} \tag{2.48}$$

for some $\gamma \in \mathbb{R}^r$, where r is the length of β. Moreover, let us define the set

$$\operatorname{cone}(h) = \{h_1 \in \Xi_{n,2m} : h_1(x) = h(x) + s(x; \gamma), \ \gamma > 0\}. \tag{2.49}$$

Then,

$$\operatorname{cone}(h) \subset \Delta_{n,2m}. \tag{2.50}$$

Moreover,

$$\exists \delta > 0 : \mu(h + s(\cdot, \gamma)) \geq \mu(h) + \delta \min_{1 \leq i \leq r} \gamma_i. \tag{2.51}$$

Proof. First of all, let us observe that $s(x; \gamma)$ is SOS for all $\gamma \geq 0$ because a positive semidefinite SMR matrix of $s(x; \gamma)$ for all $\gamma \geq 0$ is given by

$$S(\gamma) = V_p \operatorname{diag}(\gamma) V_p'.$$

In order to prove (2.50), let us observe that

$$H_1 = H + S(\gamma)$$

is a maximal SMR matrix of

$$h_1(x) = h(x) + s(x; \gamma).$$

In fact, we have that

$$
H_1 = \begin{pmatrix} V_0 & V_p \end{pmatrix} \left(\lambda_{min}(H) I_{\sigma(n,m)} + \mathrm{diag}\begin{pmatrix} 0_{\sigma(n,m)-r} \\ \gamma \end{pmatrix} \right) \begin{pmatrix} V_0' \\ V_p' \end{pmatrix} + V_p \, \mathrm{diag}(\beta) V_p'
$$

$$
= \begin{pmatrix} V_0 & V_p \end{pmatrix} \left(\lambda_{min}(H) I_{\sigma(n,m)} + \mathrm{diag}\begin{pmatrix} 0_{\sigma(n,m)-r} \\ \beta + \gamma \end{pmatrix} \right) \begin{pmatrix} V_0' \\ V_p' \end{pmatrix}
$$

which clearly implies that

$$\langle \lambda_{min}(H), \beta + \gamma, V_0, V_p \rangle \tag{2.52}$$

is a minimum eigenvalue decomposition of H_1. Since H is a maximal SMR matrix of $h(x)$, we have from Theorem 2.6 that $\eta^*(V_0) \leq 0$, which implies that also H_1 is a maximal SMR matrix.

Now, from the fact that H_1 is a maximal SMR matrix and taking into account its minimum eigenvalue decomposition in (2.52), it follows that

$$\lambda(h_1) = \lambda_{min}(H_1) = \lambda_{min}(H) = \lambda(h).$$

Moreover, we have that

$$\mu(h_1) \geq \mu(h)$$

because $s(x; \gamma)$ is SOS. Since $h \in \Delta_{n,2m}$ we conclude that $\lambda(h_1) = \lambda(h) < 0$ and $\mu(h_1) \geq \mu(h) \geq 0$, which imply that $h_1(x)$ is PNS. Therefore, (2.50) holds.

Finally, let us observe that

$$\mu(h + s(\cdot, \gamma)) \geq \mu(h) + \mu(s(\cdot, \gamma))$$

and

$$s(x; \gamma) \geq \|V_p' x^{\{m\}}\|^2 \min_{1 \leq i \leq r} \gamma_i \quad \forall x \, \forall \gamma.$$

According to Theorem 2.12, we have that $V_p' x^{\{m\}} \neq 0$ for all $x \in \mathbb{R}_0^n$. Hence, (2.51) holds with $\delta = \mu(h_2)$, where $h_2(x)$ is the form $h_2(x) = \|V_p' x^{\{m\}}\|^2$. $\qquad \square$

Theorem 2.13 states that any PNS form $h(x)$ is the vertex of a cone of PNS forms given by cone(h). In particular, the directions of this cone correspond to the SOS forms given by $s(x; \gamma)$ for $\gamma > 0$. Let us also observe that, according to (2.51), there exist PNS forms in this cone whose positivity index is arbitrarily large.

2.4.3 Parametrization of PNS Forms

Maximal SMR matrices can be exploited to derive a parametrization of all PNS forms. Let us define the set

$$\Theta^P_{n,2m}(r) = \Big\{ V_p \in \mathbb{R}^{\sigma(n,m) \times r} : \tag{2.53}$$
$$V'_p V_p = I_r, \ \eta^*(\text{cmp}(V_p)) \le 0, \text{ and (2.47) holds} \Big\}.$$

The notation $\text{cmp}(V_p)$ denotes any matrix in $\mathbb{R}^{\sigma(n,m) \times (\sigma(n,m)-r)}$ whose columns are an orthonormal base of $\ker(V'_p)$. Hence $\text{cmp}(V_p)$ satisfies the conditions

$$\begin{cases} \text{cmp}(V_p)' \text{cmp}(V_p) = I_{\sigma(n,m)-r} \\ \text{img}(\text{cmp}(V_p)) = \ker(V'_p). \end{cases}$$

Now, let us introduce the set

$$\Theta_{n,2m} = \bigcup_{1 \le r \le \sigma(n,m)} \Theta_{n,2m}(r) \tag{2.54}$$

where

$$\Theta_{n,2m}(r) = \Big\{ \langle \delta, \beta, V_p \rangle : \ \delta \in (0,1]; \ \beta \in \mathbb{R}^r, \beta > 0; \ V_p \in \Theta^P_{n,2m}(r) \Big\}. \tag{2.55}$$

For any $\theta = \langle \delta, \beta, V_p \rangle \in \Theta_{n,2m}(r)$, let $s(x; \beta) = x^{\{m\}'} V_p \text{diag}(\beta) V'_p x^{\{m\}}$ and define the form

$$\pi(x; \theta) = s(x; \beta) - \delta \mu\left(s(\cdot, \beta)\right) x^{\{m\}'} x^{\{m\}}. \tag{2.56}$$

The following result provides a parametrization of the set of PNS forms $\Delta_{n,2m}$.

Theorem 2.14. *Let $\Theta_{n,2m}$ be defined by (2.53)–(2.55), and $\pi(x; \theta)$ be given by (2.56). Then,*

$$h \in \Delta_{n,2m} \iff \exists \theta \in \Theta_{n,2m} : h(x) = \pi(x; \theta). \tag{2.57}$$

Proof. (Necessity) Let $h \in \Delta_{n,2m}$. Let H be a maximal SMR matrix of $h(x)$, and let $\langle \lambda_{min}(H), \beta, V_0, V_p \rangle$ be a minimum eigenvalue decomposition of H. Let r be the length of β. We have that

$$h(x) = x^{\{m\}'} \begin{pmatrix} V_0 & V_p \end{pmatrix} \left(\lambda_{min}(H) I_{\sigma(n,m)} + \text{diag}\begin{pmatrix} 0_{\sigma(n,m)-r} \\ \beta \end{pmatrix} \right) \begin{pmatrix} V'_0 \\ V'_p \end{pmatrix} x^{\{m\}}$$
$$= x^{\{m\}'} \left(\lambda_{min}(H) I_{\sigma(n,m)} + V_p \text{diag}(\beta) V'_p \right) x^{\{m\}}$$
$$= \lambda_{min}(H) x^{\{m\}'} x^{\{m\}} + s(x; \beta).$$

Hence, $h(x) = \pi(x; \theta)$ where

$$\theta = \langle \delta, \beta, V_p \rangle$$

$$\delta = -\frac{\lambda_{min}(H)}{\mu(s(\cdot, \beta))}.$$

Let us observe that $\delta \in (0,1]$ because $\lambda_{min}(H) = \lambda(h) < 0$ and $\lambda_{min}(H) + \mu(s(\cdot,\beta)) = \mu(g(x)) \geq 0$. Moreover, $\beta > 0$ because $\langle \lambda_{min}(H), \beta, V_0, V_p \rangle$ is a minimum eigenvalue decomposition of H. Then, by Theorem 2.6 and Theorem 2.12 it follows that $V_p \in \Theta^P_{n,2m}(r)$.

(Sufficiency) Let $\theta = \langle \delta, \beta, V_p \rangle \in \Theta_{n,2m}$. We have that an SMR matrix of $\pi(x;\theta)$ is given by

$$H = V_p \text{diag}(\beta)V_p' - \delta\mu(s(\cdot,\beta))I_{\sigma(n,m)}$$

$$= \left(\text{cmp}(V_p) \quad V_p \right) \left(\text{diag} \begin{pmatrix} 0_{\sigma(n,m)-r} \\ \beta \end{pmatrix} - \delta\mu(s(\cdot,\beta))I_{\sigma(n,m)} \right) \begin{pmatrix} \text{cmp}(V_p)' \\ V_p' \end{pmatrix}.$$

Since $V_p'V_p = I_r$ and $\beta > 0$, it follows that

$$\langle -\delta\mu(s(\cdot,\beta)), \beta, \text{cmp}(V_p), V_p \rangle$$

is a minimum eigenvalue decomposition of H. Since $\eta^*(\text{cmp}(V_p)) \leq 0$, this implies that H is a maximal SMR matrix from Theorem 2.6. Moreover, from Theorem 2.12 it follows that $\mu(s(\cdot,\beta)) > 0$. Hence,

$$\lambda(\pi(\cdot,\theta)) = -\delta\mu(s(\cdot,\beta)) < 0$$

and

$$\mu(\pi(\cdot,\theta)) = (1-\delta)\mu(s(\cdot,\beta)) \geq 0.$$

Therefore, (2.57) holds.   □

Theorem 2.14 states that $\Delta_{n,2m}$ is the image of $\Theta_{n,2m}$ through the function $\pi(x;\theta)$. Hence, this result provides a technique to parametrize and construct all the PNS forms. This technique amounts to finding matrices V_p in $\Theta^P_{n,2m}(r)$ and calculating the positivity index $\mu(s(\cdot,\beta))$. Unfortunately, it is difficult to find an explicit representation of the set $\Theta^P_{n,2m}(r)$. A method to find elements in this set consists of looking for matrices V_p with a fixed structure, for which the property (2.47) and the positivity index $\mu(s(\cdot,\beta))$ can be easily checked, and using the remaining free parameters to satisfy the condition $\eta^*(\text{cmp}(V_p)) \leq 0$.

Example 2.10. We show here the construction of a simple PNS by using Theorem 2.14, in the case with $n = 3$ and $m = 3$. Let us choose $x^{\{m\}}$ as

$$x^{\{m\}} = (x_1^3, \sqrt{3}x_1^2x_2, \sqrt{3}x_1^2x_3, \sqrt{3}x_1x_2^2\sqrt{6}x_1x_2x_3, \sqrt{3}x_1x_3^2, x_2^3, \tag{2.58}$$
$$\sqrt{3}x_2^2x_3, \sqrt{3}x_2x_3^2, x_3^3)'.$$

This choice satisfies (1.9). Then, let us choose

$$V_p = \frac{1}{\sqrt{38}} \begin{pmatrix} 2\sqrt{3} & 0 & 0 & -5 & 0 & 1 & 0 & 0 & 0 & 0 \\ 0 & 1 & 0 & 0 & 0 & 0 & 2\sqrt{3} & 0 & -5 & 0 \\ 0 & 0 & -5 & 0 & 0 & 0 & 0 & 1 & 0 & 2\sqrt{3} \end{pmatrix}'.$$

The number of columns of V_p is $r = 3$. Observe that $V_p'V_p = I_3$. Moreover, by selecting $y = (1,0,\ldots,0)'$ in (2.27), we find that $\eta\,(\mathrm{cmp}(V_p)) = -0.0792$, which allows us to conclude that $\eta^*(\mathrm{cmp}(V_p)) \le 0$, by Theorem 2.7. Thanks to the structure of V_p, it is easy to verify the property (2.47) and to compute the positivity index $\mu\,(s(\cdot,\beta))$. In fact,

$$V_p' x^{\{m\}} = \sqrt{\frac{3}{38}}\,(w_1(x), w_2(x), w_3(x))'$$

where

$$w_1(x) = x_1\left(2x_1^2 - 5x_2^2 + x_3^2\right)$$
$$w_2(x) = x_2\left(x_1^2 + 2x_2^2 - 5x_3^2\right)$$
$$w_3(x) = x_3\left(-5x_1^2 + x_2^2 + 2x_3^2\right).$$

It is straightforward to see that

$$w_i(x) = 0 \;\forall i = 1,2,3 \iff x = 0_3$$

and hence (2.47) holds. Therefore, $V_p \in \Theta_{3,6}^P(3)$ and

$$\theta = \langle \delta, \beta, V_p \rangle \in \Theta_{3,6} \quad \forall \delta \in (0,1], \forall \beta \in \mathbb{R}^3, \beta > 0.$$

Moreover, let us select a vector β, for example $\beta = (38/3, 38/3, 38/3)'$. It follows

$$s(x;\beta) = \sum_{i=1}^{3} w_i(x)^2. \tag{2.59}$$

In order to compute $\mu\,(s(\cdot,\beta))$, we have to find the minimum of $s(x;\beta)$ over the set $\mathscr{C}_{n,m}$, which coincides with $\{x : \|x\| = 1\}$ due to the choice (2.58). Let us observe that, since $s(x;\beta)$ depends on x_1^2, x_2^2, x_3^2, one can first substitute $x_3^2 = 1 - x_1^2 - x_2^2$ in $s(x;\beta)$, and then find the minimum by computing the points where the derivatives of $s(x;\beta)$ with respect to x_1^2 and x_2^2 vanish. This operation amounts to solving a system of two quadratic equations in two variables, and can be done by finding the roots of a polynomial equation of degree four in one variable. We find

$$\mu\,(s(\cdot,\beta)) = 0.4360.$$

Let us define

$$h_0(x) = \sum_{i=1}^{3} w_i(x)^2$$
$$= 4(x_1^6 + x_2^6 + x_3^6) - 19(x_1^4 x_2^2 + x_2^4 x_3^2 + x_3^4 x_1^2) + 29(x_1^4 x_3^2 + x_2^4 x_1^2 + x_3^4 x_2^2)$$
$$- 30 x_1^2 x_2^2 x_3^2.$$

Therefore, from Theorem 2.14 it follows that the form

$$h(x) = h_0(x) - 0.4360\|x\|^6 \delta \tag{2.60}$$

is a PNS form for all $\delta \in (0,1]$. Figure 2.1 shows the plot of $h(x)$ on the upper semi-sphere for $\delta = 0.5$.

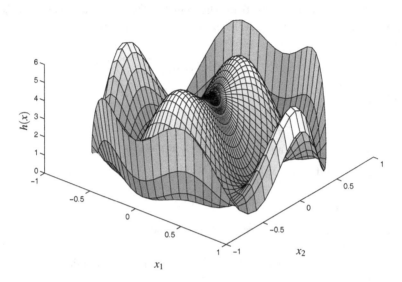

Fig. 2.1 Example 2.10: form $h(x)$ in (2.60) plotted with $\delta = 0.5$ for x such that $x_1^2 + x_2^2 \leq 1$ and $x_3 = \sqrt{1 - x_1^2 - x_2^2}$

Example 2.11. Let us consider the PNS form in (2.60), and let $\gamma \in \mathbb{R}^3$. From (2.59) we have that

$$h(x) + s(x;\gamma) = \sum_{i=1}^{3}(1+\gamma_i)w_i(x)^2 - 0.4360\|x\|^6\delta.$$

This implies that the cone (2.49) is given by

$$\text{cone}(h) = \left\{ h_1 \in \Xi_{3,6} : h_1(x) = \sum_{i=1}^{3}(1+\gamma_i)w_i(x)^2 - 0.4360\|x\|^6\delta, \ \gamma \geq 0 \right\}.$$

According to Theorem 2.13, such a cone contains only PNS forms.

2.5   Notes and References

There is an endless literature on Hilbert's 17th problem and related issues. The interested reader is referred to the classical book [70], and to more recent contributions such as [118, 122, 123] and references therein.

Theorem 2.4 was given in [34]. Maximal SMR matrices and related results in Section 2.2, as well as the characterization of PNS forms in Section 2.4, have been provided in [25]. SMR-tight forms have been introduced in [27].

The study of the gap between positive forms and SOS forms is a classical problem which has recently attracted much interest, see e.g. [9, 86, 25].

Chapter 3
Robustness with Time-varying Uncertainty

This chapter addresses robust stability of time-varying systems affected by structured parametric uncertainty, a fundamental problem in robust control. It is shown that the problem can be tackled by employing Lyapunov functions which are forms in the state variables and are referred as HPLFs. Thanks to the tools for checking positivity of forms introduced in Chapters 1 and 2, the construction of such Lyapunov functions can be formulated in terms of special convex optimizations problems. Both polytopic systems and LFRs are considered. Several robustness performance measures are also investigated.

3.1 Polytopic Systems with Time-varying Uncertainty

Let us start by introducing linear systems affected by time-varying structured parametric uncertainty.

Definition 3.1 (Time-varying Polytopic System). Consider the continuous-time system described by the state equations

$$\dot{x}(t) = A(p(t))x(t) \tag{3.1}$$

where $t \in \mathbb{R}$ is the time, $x(t) \in \mathbb{R}^n$ is the state vector, $\dot{x}(t) = \frac{dx(t)}{dt}$, $p(t) \in \mathbb{R}^q$ is an uncertain parameter vector, and $A(p(t)) \in \mathbb{R}^{n \times n}$ is given by

$$A(p(t)) = A_0 + \sum_{i=1}^{q} p_i(t)A_i \tag{3.2}$$

where $A_0, \ldots, A_q \in \mathbb{R}^{n \times n}$ are given matrices. It is assumed that $p(t)$ satisfies the constraint

$$p(t) \in \mathscr{P}, \quad \forall t \geq 0 \tag{3.3}$$

being $\mathscr{P} \subset \mathbb{R}^q$ the polytope defined as

G. Chesi et al.: Homogeneous Polynomial Forms, LNCIS 390, pp. 63–97.
springerlink.com   © Springer-Verlag Berlin Heidelberg 2009

$$\mathscr{P} = \mathrm{co}\{p^{(1)}, \dots, p^{(r)}\} \tag{3.4}$$

for some given vectors $p^{(1)}, \dots, p^{(r)} \in \mathbb{R}^q$. Then, the system (3.1)–(3.4) is called *time-varying polytopic system*.

The vector $p(t)$ represents the time-varying parametric uncertainty which affects affinely the system dynamics. This vector can be any piecewise continuous function of time t, provided that $p(t) \in \mathscr{P}$. The set $\mathscr{P}$ is the polytope described by the convex hull of the vectors $p^{(1)}, \dots, p^{(r)}$.

A fundamental problem for time-varying polytopic systems is to establish whether the following property holds.

Definition 3.2 (Robust Stability for Time-varying Polytopic System). The system (3.1)–(3.4) is said *robustly stable* if the following conditions hold:

1. $\forall \varepsilon > 0 \ \exists \delta > 0 : \|x(0)\| < \delta \Rightarrow \|x(t)\| \le \varepsilon \ \ \forall t \ge 0, \ \forall p(t) \in \mathscr{P}$;
2. $\lim_{t \to \infty} x(t) = 0_n \ \ \forall x(0) \in \mathbb{R}^n, \ \forall p(t) \in \mathscr{P}$.

According to the previous definition, the system (3.1) is robustly stable whenever its origin is a globally asymptotically stable equilibrium point for all possible parametric uncertainties satisfying (3.3)–(3.4).

In the sequel the dependence on the time t will be omitted for ease of presentation, unless it is required by the context.

3.1.1 Homogeneous Polynomial Lyapunov Functions

In the following, we will show that robustness issues relevant to time-varying polytopic systems such as the robust stability property in Definition 3.2 can be investigated by using Lyapunov functions which are forms in the state variables. The next definition introduces this class of Lyapunov functions.

Definition 3.3 (HPLF). Let $v : \mathbb{R}^n \to \mathbb{R}$ be a function satisfying

$$\begin{cases} v \in \varXi_{n,2m} \\ v(x) > 0 \ \ \forall x \in \mathbb{R}_0^n \\ \dot{v}(x) < 0 \ \ \forall x \in \mathbb{R}_0^n \ \forall p \in \mathscr{P} \end{cases} \tag{3.5}$$

where

$$\dot{v}(x) = \left. \frac{dv(x)}{dt} \right|_{\dot{x}=A(p)x} . \tag{3.6}$$

Then, $v(x)$ is a *HPLF* of degree $2m$ for the system (3.1)–(3.4).

Hence, HPLFs are forms allowing one to prove robust global asymptotic stability of the origin of (3.1)–(3.4). These Lyapunov functions can be represented via the SMR according to (1.11), as

$$v(x) = x^{\{m\}'} V x^{\{m\}} \tag{3.7}$$

where $V \in \mathbb{S}^{\sigma(n,m)}$.

In the next sections it will be explained how to verify that a given form is an HPLF and how to construct HPLFs by solving convex optimization problems.

3.1.2 Extended Matrix

In order to address robustness analysis via HPLFs, we first need to introduce the notion of extended matrix.

Definition 3.4 (Extended Matrix). Let us consider the system

$$\dot{x} = Ax$$

where $x \in \mathbb{R}^n$ and $A \in \mathbb{R}^{n \times n}$. Let $m \geq 1$ be an integer, and let $A^{\#} \in \mathbb{R}^{\sigma(n,m) \times \sigma(n,m)}$ be the matrix satisfying the relation

$$\begin{aligned} \frac{dx^{\{m\}}}{dt} &= \frac{\partial x^{\{m\}}}{\partial x} Ax \\ &= A^{\#} x^{\{m\}}. \end{aligned} \tag{3.8}$$

The matrix $A^{\#}$ is called *extended matrix* of A.

An expression of the extended matrix is provided below in terms of Kronecker's products.

Theorem 3.1. *Let $K_m \in \mathbb{R}^{n^m \times \sigma(n,m)}$ be the matrix satisfying*

$$x^{[m]} = K_m x^{\{m\}} \tag{3.9}$$

where $x^{[m]}$ denotes the m-th Kronecker power of x. Then, $A^{\#}$ is given by

$$A^{\#} = (K_m' K_m)^{-1} K_m' \left(\sum_{i=0}^{m-1} I_{n^{m-1-i}} \otimes A \otimes I_{n^i} \right) K_m. \tag{3.10}$$

Proof. From (3.8) and (3.9) it follows that

$$\begin{aligned} K_m A^{\#} x^{\{m\}} &= \frac{\partial x^{[m]}}{\partial x} Ax \\ &= \left(\sum_{j=0}^{m-1} x^{[j]} \otimes I_n \otimes x^{[m-1-j]} \right) Ax. \end{aligned}$$

Then, for any $j = 0, \ldots, m-1$ one has that

$$\left(x^{[j]} \otimes I_n \otimes x^{[m-1-j]}\right) Ax = x^{[j]} \otimes Ax \otimes x^{[m-1-j]}$$
$$= \left(I_{n^j} \otimes A \otimes I_{n^{m-1-j}}\right) x^{[m]}$$
$$= \left(I_{n^j} \otimes A \otimes I_{n^{m-1-j}}\right) K_m x^{\{m\}}$$

and hence (3.10) holds.   □

Example 3.1. Let us consider

$$A = \begin{pmatrix} -1 & 1 \\ -2 & -1 \end{pmatrix}.$$

Then, for $m = 2$ with $x^{\{2\}} = (x_1^2, x_1 x_2, x_2^2)'$, one obtains from (3.8) the extended matrix

$$A^{\#} = \begin{pmatrix} -2 & 2 & 0 \\ -2 & -2 & 1 \\ 0 & -4 & -2 \end{pmatrix}$$

whereas for $m = 3$, with $x^{\{3\}} = (x_1^3, x_1^2 x_2, x_1 x_2^2, x_2^3)'$, one has

$$A^{\#} = \begin{pmatrix} -3 & 3 & 0 & 0 \\ -2 & -3 & 2 & 0 \\ 0 & -4 & -3 & 1 \\ 0 & 0 & -6 & -3 \end{pmatrix}.$$

The following result provides a key property of the extended matrix $A^{\#}$ that will be exploited in the sequel.

Theorem 3.2. *For $i = 0, \ldots, q$ let $A_i^{\#}$ denote the extended matrix of A_i, and let $A(p)^{\#}$ denote the extended matrix of $A(p)$. Then,*

$$A(p)^{\#} = A_0^{\#} + \sum_{i=1}^{q} p_i A_i^{\#} \quad \forall p \in \mathbb{R}^q. \tag{3.11}$$

Proof. From the definition of the extended matrix in (3.8) it follows that $A^{\#}$ depends linearly on A, and hence (3.11) holds.   □

3.2 Robust Stability

This section investigates robust stability of the system (3.1)–(3.4) by using HPLFs and the results derived in Chapters 1 and 2.

Let us start by showing that a sufficient condition for establishing the existence of an HPLF can be obtained through the SOS index introduced in (1.29).

Theorem 3.3. *Let $v \in \Xi_{n,2m}$. Assume that*

$$\begin{cases} 0 < \lambda(v) \\ 0 < \lambda(-d_i) & \forall i = 1, \ldots, r \end{cases} \tag{3.12}$$

where $d_i \in \Xi_{n,2m}$ is defined as

$$d_i(x) = \dot{v}(x)|_{p=p^{(i)}}. \tag{3.13}$$

Then, $v(x)$ is an HPLF for the system (3.1)–(3.4).

Proof. Let us suppose that (3.12) holds. Then, from Corollary 1.4 one has that $v(x)$ is positive definite and $d_1(x), \ldots, d_r(x)$ are negative definite. Now, from (3.11) one has that

$$\dot{v}(x) = \sum_{i=1}^{r} c_i(p) d_i(x)$$

where $c_1(p), \ldots, c_r(p) \in \mathbb{R}$ are the coefficients which allow one to express $p \in \mathscr{P}$ as a convex combination of the vertices of $\mathscr{P}$, i.e.

$$\begin{cases} \displaystyle\sum_{i=1}^{r} c_i(p) p^{(i)} = p \\ \displaystyle\sum_{i=1}^{r} c_i(p) = 1 \\ c_i(p) \geq 0 \quad \forall i = 1, \ldots, r. \end{cases}$$

In particular, $\forall p \in \mathscr{P}$ there exists at least one integer i, $1 \leq i \leq r$, such that $c_i(p) > 0$, and hence $\dot{v}(x) < 0$ for all $x \in \mathbb{R}_0^n$ and for all $p \in \mathscr{P}$. Therefore, (3.5) holds, i.e. $v(x)$ is an HPLF for the system (3.1)–(3.4). $\qquad\square$

Theorem 3.3 can be used for constructing an HPLF by solving an LMI feasibility test. Indeed, for $i = 1, \ldots, r$ let us define the matrices

$$\hat{A}_i = A(p^{(i)})$$

and let $\hat{A}_i^{\#}$ denote the extended matrix of $\hat{A}_i$.

Theorem 3.4. *Let $m \geq 1$ be an integer, and $L(\cdot)$ be a linear parametrization of $\mathscr{L}_{n,m}$ in (1.15). Let us suppose that there exist $V \in \mathbb{S}^{\sigma(n,m)}$ and $\alpha^{(1)}, \ldots, \alpha^{(r)} \in \mathbb{R}^{\omega(n,m)}$ satisfying the following system of LMIs:*

$$\begin{cases} 0 < V \\ 0 > \text{he}\left(V\hat{A}_i^{\#}\right) + L(\alpha^{(i)}), & i = 1, \ldots, r. \end{cases} \tag{3.14}$$

Then, $v(x)$ defined as in (3.7) is an HPLF of degree $2m$ for the system (3.1)–(3.4).

Proof. By (3.13) and (3.8), one has

$$d_i(x) = 2x^{\{m\}'}V\frac{\partial x^{\{m\}}}{\partial x}\hat{A}_i x$$
$$= x^{\{m\}'}\text{he}\left(V\hat{A}_i^{\#}\right)x^{\{m\}}.$$

Therefore, the existence of $V, \alpha^{(1)}, \ldots, \alpha^{(r)}$ such that (3.14) holds, is equivalent to condition (3.12).   □

Notice that the scalar variables in (3.14) are the elements of the matrix V and vectors $\alpha^{(1)}, \ldots, \alpha^{(r)}$, and therefore their number is equal to

$$\frac{1}{2}\sigma(n,m)(\sigma(n,m)+1)+r\omega(n,m).$$

Table 3.1 shows this number for some values of n, m, r. Let us observe that $V, \alpha^{(1)}, \ldots, \alpha^{(r)}$ are defined up to a positive scalar factor since the inequalities in (3.14) are linear in these variables.

Table 3.1 Number of scalar variables in the LMI feasibility test (3.14) for some values of n, m, r: (a) $r = 2$; (b) $r = 3$

	(a)				(b)			
	$m=1$	$m=2$	$m=3$	$m=4$	$m=1$	$m=2$	$m=3$	$m=4$
$n=1$	1	1	1	1	1	1	1	1
$n=2$	3	8	16	27	3	9	19	33
$n=3$	6	33	109	270	6	39	136	345
$n=4$	10	95	462	1560	10	115	588	2025

A natural question that arises is whether the conservatism of the sufficient condition provided by Theorem 3.4 is nonincreasing with the degree of the HPLF. The following result provides an answer to this question.

Theorem 3.5. *If condition (3.14) of Theorem 3.4 holds for some integer $m \geq 1$, then it holds also for km where k is any integer satisfying $k \geq 1$.*

Proof. Let $V, \alpha^{(1)}, \ldots, \alpha^{(r)}$ be such that (3.14) holds for a given m, and consider any integer $k \geq 1$. We now show that there exist $\tilde{V}, \tilde{\alpha}^{(1)}, \ldots, \tilde{\alpha}^{(r)}$ such that (3.14) holds with m replaced by km. Let us define $v(x)$ as in (3.7). We have that (3.5) holds. Then, let us introduce the form

$$\tilde{v}(x) = v(x)^k.$$

We have that (3.5) holds with $v(x)$ and m replaced by $\tilde{v}(x)$ and km respectively, i.e. $\tilde{v}(x)$ is an HPLF of degree $2km$ for the system (3.1)–(3.4). Now, let us define

$$\tilde{V} = T'V^{[k]}T \tag{3.15}$$

where T is the matrix satisfying

$$\left(x^{\{m\}}\right)^{[k]} = Tx^{\{km\}} \quad \forall x.$$

We have that

$$\tilde{v}(x) = x^{\{km\}'}\tilde{V}x^{\{km\}}.$$

Being $V > 0$ and T full column rank, from (3.15) one has $\tilde{V} > 0$. Then, let us define

$$W_i = T'\left(V^{[k-1]} \otimes \left(\text{he}\left(V\hat{A}_i^{\#}\right) + L(\alpha^{(i)})\right)\right)T$$

for $i = 1,\ldots,r$. We have that

$$\left.\frac{d\tilde{v}(x)}{dt}\right|_{\dot{x}=\hat{A}_i x} = x^{\{km\}'}W_i x^{\{km\}}$$

and by (3.14), $W_i < 0$. By exploiting (3.8), one has

$$\left.\frac{d\tilde{v}(x)}{dt}\right|_{\dot{x}=\hat{A}_i x} = x^{\{km\}'}\,\text{he}\left(\tilde{V}X_i\right)x^{\{km\}}$$

where X_i is the extended matrix of $\hat{A}_i$ defined by (3.8), with m replaced by km. Hence, there exist $\tilde{\alpha}^{(i)}$, $i = 1,\ldots,r$, such that

$$\text{he}\left(\tilde{V}X_i\right) + L(\tilde{\alpha}^{(i)}) = W_i$$

because he $\left(\tilde{V}X_i\right)$ and W_i are SMR matrices of the same form. $\qquad\square$

For some special values of the dimension n of x and the degree $2m$ of $v(x)$, the sufficient condition provided by Theorem 3.4 is also necessary for the existence of an HPLF for the system (3.1)–(3.4). This is explained by the following result.

Theorem 3.6. *Let us suppose that $(n,2m)$ belongs to $\mathscr{E}$ in (2.5). Then, there exists an HPLF of degree $2m$ for the system (3.1)–(3.4) if and only if there exist $V, \alpha^{(1)},\ldots,\alpha^{(r)}$ such that (3.14) holds.*

Proof. Obviously, it must be proven only that if $(n,2m) \in \mathscr{E}$ and there exists an HPLF $v(x)$ of degree $2m$ for (3.1)–(3.4), then (3.14) admits a feasible solution. By assumption, we have that $v(x)$ is positive definite and $-\dot{v}(x)$ is positive definite for all $p \in \mathscr{P}$. Hence, due to Theorem 2.4, there exist $V, V_1,\ldots,V_r$ such that

$$\begin{cases} 0 < V \\ 0 < V_i, \quad i = 1, \ldots, r \\ v(x) = x^{\{m\}'} V x^{\{m\}} \\ -d_i(x) = x^{\{m\}'} V_i x^{\{m\}}, \quad i = 1, \ldots, r. \end{cases}$$

Completeness of the SMR parametrization (1.17) implies that there exist $\alpha^{(1)}, \ldots,$ $\alpha^{(r)}$ such that

$$-V_i = \text{he}\left(V \hat{A}_i^\#\right) + L(\alpha^{(i)}), \quad i = 1, \ldots, r.$$

Therefore, (3.14) admits a feasible solution.   □

A general converse theorem for HPLFs has been proven in [8] and is reported next. Unfortunately, the degree of the resulting HPLF can be arbitrarily high.

Theorem 3.7. *If the system (3.1)–(3.4) is robustly stable, then there exists an integer $m \geq 1$ and a matrix $V \in \mathbb{S}^{\sigma(n,m)}$ such that $v(x) = x^{\{m\}'} V x^{\{m\}}$ is an HPLF for the system (3.1)–(3.4).*

Proof. It is a direct consequence of Theorem 3.2 in [8].   □

Example 3.2. Let us consider the system (3.1)–(3.4) with $x \in \mathbb{R}^2$, $p \in \mathbb{R}$ and

$$A_0 = \begin{pmatrix} -1 & 1 \\ -2 & -1 \end{pmatrix}, \quad A_1 = \begin{pmatrix} 0 & 0 \\ 1 & 0 \end{pmatrix}, \quad p^{(1)} = 0, \quad p^{(2)} = 1.$$

Let us define the function

$$v(x) = x_1^4 + x_2^4.$$

Then, $v(x)$ is an HPLF of degree 4 for this system. Indeed, (3.14) holds with

$$V = \begin{pmatrix} 1 & 0 & -0.5 \\ \star & 1 & 0 \\ \star & \star & 1 \end{pmatrix}, \quad \alpha^{(1)} = -0.7165, \quad \alpha^{(2)} = -0.2003.$$

3.3 Robust Performance

This section investigates some robust performance properties of the system (3.1), specifically the ℓ_∞ stability margin and the best transient performance index.

3.3.1 Polytopic Stability Margin

In the analysis of the time-varying polytopic system (3.1), a key problem consists of finding the largest value of a positive scalar γ for which the system is robustly stable for all uncertainties belonging to the scaled perturbation set $\gamma\mathscr{P}$, which is defined as

$$\gamma\mathscr{P} = \mathrm{co}\left\{\gamma p^{(1)}, \ldots, \gamma p^{(r)}\right\}.$$

Hence, the problem is formulated as the computation of the following robust stability margin [143, 5].

Definition 3.5 (Polytopic Stability Margin). Let us define

$$\gamma^P = \sup\ \{\gamma \in \mathbb{R} : (3.1) \text{ is robustly stable for all } p(t) \in \gamma\mathscr{P}\}.$$

Then, γ^P is called *polytopic stability margin* for the system (3.1).

In the following, we will investigate the polytopic stability margin via HPLFs. In particular, we define the polytopic stability margin guaranteed by the class of HPLFs of degree $2m$ as follows.

Definition 3.6 ($2m$-HPLF Polytopic Stability Margin). Let us define

$$\gamma_{2m}^P = \sup\ \{\gamma \in \mathbb{R} : \exists v \in \varXi_{n,2m} \text{ HPLF for (3.1) with } p(t) \in \gamma\mathscr{P}\}. \tag{3.16}$$

Then, γ_{2m}^P is called *$2m$-HPLF polytopic stability margin* for the system (3.1).

Clearly, the $2m$-HPLF polytopic stability margin is a lower bound of the sought polytopic stability margin, indeed

$$\gamma_{2m}^P \leq \gamma^P \quad \forall m \geq 1.$$

A special instance of the problem formulated above occurs when the polytope $\mathscr{P}$ is the unit ℓ_∞ box, i.e. $\mathscr{P} = \mathscr{B}_\infty(1)$ where

$$\mathscr{B}_\infty(\gamma) = \{p \in \mathbb{R}^q : \|p\|_\infty \leq \gamma\}. \tag{3.17}$$

In this case, the stability margins in Definitions 3.5 and 3.6 are referred to as *ℓ_∞ stability margin* and *$2m$-HPLF ℓ_∞ stability margin*, and denoted by γ^∞ and γ_{2m}^∞, respectively.

For ease of presentation, we address the problem of estimating γ_{2m}^∞ (the results given next are immediately extended to γ_{2m}^P). Let us denote by $u^{(1)}, \ldots, u^{(2^q)}$ the vertices of the unit ℓ_∞ ball $\mathscr{B}_\infty(1)$, introduce the matrices

$$\tilde{A}_i = A(u^{(i)}) - A_0, \quad i = 1, \ldots, 2^q$$

and let $\tilde{A}_i^\#$, $i = 0, \ldots, 2^q$, denote the extended matrix of $\tilde{A}_i$. The following result shows that a lower bound to the $2m$-HPLF ℓ_∞ stability margin γ_{2m}^∞ can be computed by solving a quasi-convex optimization problem.

Theorem 3.8. *Let $m \geq 1$ be an integer, and $L(\cdot)$ be a linear parametrization of $\mathscr{L}_{n,m}$ in (1.15). Let us define*

$$\hat{\gamma}_{2m}^{\infty} = \frac{1}{z^*} \tag{3.18}$$

where z^* is the solution of

$$
z^* = \inf_{z \in \mathbb{R},\ V \in \mathbb{S}^{\sigma(n,m)},\ \alpha^{(0)},...,\alpha^{(2^q)} \in \mathbb{R}^{\omega(n,m)}} z
$$

$$
s.t. \begin{cases} 0 < z \\ 0 < V \\ 0 < -\mathrm{he}\left(VA_0^\#\right) - L(\alpha^{(0)}) \\ 0 < z\left(-\mathrm{he}\left(VA_0^\#\right) - L(\alpha^{(0)})\right) - \mathrm{he}\left(V\tilde{A}_i^\#\right) - L(\alpha^{(i)}), \\ \quad i = 1,...,2^q. \end{cases} \tag{3.19}
$$

Then, $\hat{\gamma}_{2m}^{\infty} \leq \gamma_{2m}^{\infty}$.

Proof. The second constraint in (3.19) provides that $v(x) = x^{\{m\}'} V x^{\{m\}}$ is positive definite. The time derivative of $v(x)$ evaluated for $p = z^{-1}u^{(i)}$ is given by

$$
\begin{aligned}
\dot{v}(x)|_{p=z^{-1}u^{(i)}} &= x^{\{m\}'} \mathrm{he}\left(V\left(A_0^\# + z^{-1}\tilde{A}_i^\#\right)\right) x^{\{m\}} \\
&= z^{-1} x^{\{m\}'} \left(z\,\mathrm{he}\left(VA_0^\#\right) + \mathrm{he}\left(V\tilde{A}_i^\#\right)\right) x^{\{m\}}.
\end{aligned}
$$

Let us observe that, for all $\alpha^{(0)}, \alpha^{(i)} \in \mathbb{R}^{\omega(n,m)}$, one has that

$$
\dot{v}(x)|_{p=z^{-1}u^{(i)}} = z^{-1} x^{\{m\}'} \left(z\left(\mathrm{he}\left(VA_0^\#\right) + L(\alpha^{(0)})\right) + \mathrm{he}\left(V\tilde{A}_i^\#\right) + L(\alpha^{(i)})\right) x^{\{m\}}
$$

due to the definition of $\mathscr{L}_{n,m}$ in (1.15). Hence, the last constraint in (3.19) guarantees that

$$
\dot{v}(x)|_{p=z^{-1}u^{(i)}} \text{ is negative definite for all } i = 1,...,2^q.
$$

This clearly implies that $\dot{v}(x)$ is negative definite for all $p \in \mathscr{B}_\infty(z^{-1})$. Hence, $\hat{\gamma}_{2m}^{\infty} \leq \gamma_{2m}^{\infty}$. $\qquad\square$

Problem (3.19) is a GEVP, which belongs to the class of quasi-convex optimization problems (see Appendix A.1). Let us observe that the third LMI constraint in (3.19) guarantees that the matrix multiplying the generalized eigenvalue z is positive definite, which ensures quasi-convexity. The number of scalar variables in the GEVP (3.19) is equal to

$$
\frac{1}{2}\sigma(n,m)(\sigma(n,m)+1) + (2^q+1)\omega(n,m).
$$

The lower bound provided by Theorem 3.8 can be guaranteed to be tight *a priori*, for special values of n, m. This is explained in the following result.

Theorem 3.9. *Let us suppose that* $(n, 2m)$ *belongs to* $\mathcal{E}$ *in* (2.5). *Then,* $\hat{\gamma}_{2m}^{\infty}$ *defined by* (3.18)–(3.19) *satisfies* $\hat{\gamma}_{2m}^{\infty} = \gamma_{2m}^{\infty}$.

Proof. From (3.16) it follows that, for any z such that $z^{-1} < \gamma_{2m}^{\infty}$, there exists a form $v(x)$ of degree $2m$ in x such that $v(x)$ is positive definite and $-\dot{v}(x)$ is positive definite for all $p \in \mathcal{B}_{\infty}(z^{-1})$. Let us suppose that $(n, 2m) \in \mathcal{E}$. From Theorem 2.4 it follows that $v(x)$ and $-\dot{v}(x)$ admit positive definite SMR matrices for any $p \in \mathcal{B}_{\infty}(z^{-1})$. Hence, the constraint in problem (3.19) admits a feasible solution for any $z^{-1} < \gamma_{2m}^{\infty}$. Therefore, $\hat{\gamma}_{2m}^{\infty} = \gamma_{2m}^{\infty}$. □

A simpler version of Theorem 3.8 can be obtained for the special case of a segment of matrices. Since the problem is widely addressed in the literature, and will be considered in several examples in Section 3.5, we work out the details. Consider the system

$$\begin{cases} \dot{x} = (A_0 + pA_1)x \\ p \in [0, \kappa]. \end{cases} \tag{3.20}$$

The aim is to compute the largest value of κ for which the system (3.20) is robustly stable, i.e.

$$\kappa^{\infty} = \sup \left\{ \kappa \in \mathbb{R} : (3.20) \text{ is robustly stable for all } p(t) \in [0, \kappa] \right\}. \tag{3.21}$$

Clearly, one can define κ_{2m}^{∞} as the largest value of κ for which there exists an HPLF of degree $2m$ for the system (3.20). Then, the following result easily follows from Theorems 3.8 and 3.9.

Corollary 3.1. *Let* $m \geq 1$ *be an integer, and* $L(\cdot)$ *be a linear parametrization of* $\mathcal{L}_{n,m}$ *in* (1.15). *Let us define*

$$\hat{\kappa}_{2m}^{\infty} = \frac{1}{z^*}$$

where

$$z^* = \inf_{z \in \mathbb{R},\, V \in \mathcal{S}^{\sigma(n,m)},\, \alpha^{(1)}, \alpha^{(2)} \in \mathbb{R}^{\omega(n,m)}} z$$

$$s.t. \begin{cases} 0 < V \\ 0 < -\mathrm{he}\left(VA_0^{\#}\right) - L(\alpha^{(1)}) \\ 0 < z\left(-\mathrm{he}\left(VA_0^{\#}\right) - L(\alpha^{(1)})\right) - \mathrm{he}\left(VA_1^{\#}\right) + L(\alpha^{(2)}). \end{cases} \tag{3.22}$$

Then, $\hat{\kappa}_{2m}^{\infty} \leq \kappa_{2m}^{\infty}$. *Moreover, if* $(n, 2m)$ *belongs to* $\mathcal{E}$ *in* (2.5)*, then* $\hat{\kappa}_{2m}^{\infty} = \kappa_{2m}^{\infty}$.

3.3.2 Best Transient Performance

Another problem of interest in robustness analysis of uncertain systems consists of determining the Lyapunov function that achieves the best transient performance. For a given Lyapunov function $v(x)$ proving robust stability of the system (3.1)–(3.4), one can define the transient performance index as

$$\gamma^T(v) = \sup_{x \in \mathbb{R}_0^n} \sup_{p \in \mathscr{P}} \frac{\dot{v}(x)}{v(x)}. \tag{3.23}$$

From (3.23) one has

$$v(x(t)) \leq v(x(t_0)) e^{\gamma^T(v)(t-t_0)}$$

thus establishing the rate of decrease of $v(x)$. Therefore, it is natural to select among all feasible Lyapunov functions the one that minimizes $\gamma^T(v)$, in order to obtain the fastest transient. We introduce the following definition.

Definition 3.7 (Best Transient Performance Index). Let us define

$$\gamma^T = \inf_v \gamma^T(v). \tag{3.24}$$

Then, γ^T is called *best transient performance index* for the system (3.1)–(3.4).

In this section the aim is to investigate the best transient performance index via HPLFs. We hence introduce the following definition.

Definition 3.8 (2m-HPLF Best Transient Performance Index). Let us define

$$\gamma_{2m}^T = \inf_{v \in \Xi_{n,2m}} \gamma^T(v). \tag{3.25}$$

Then, γ_{2m}^T is called *2m-HPLF best transient performance index* for the system (3.1)–(3.4).

Clearly, the $2m$-HPLF best transient performance index provides an upper bound to the sought best transient performance index, indeed

$$\gamma_{2m}^T \geq \gamma^T \quad \forall m \geq 1.$$

The following result shows that an upper bound to γ_{2m}^T can be obtained by solving a GEVP.

Theorem 3.10. *Let $m \geq 1$ be an integer, and $L(\cdot)$ be a linear parametrization of $\mathscr{L}_{n,m}$ in (1.15). Let us define the GEVP*

$$\hat{\gamma}_{2m}^T = \inf_{z \in \mathbb{R}, \, V \in \mathbb{S}^{\sigma(n,m)}, \, \alpha^{(i)} \in \mathbb{R}^{\omega(n,m)}, \, i=1,\dots,r} z$$

$$s.t. \begin{cases} 0 < V \\ 0 < zV - \mathrm{he}\left(V\hat{A}_i^\#\right) - L(\alpha^{(i)}), \quad i=1,\dots,r. \end{cases} \tag{3.26}$$

Then, $\hat{\gamma}_{2m}^T \geq \gamma_{2m}^T$. Moreover, if $(n,2m)$ belongs to $\mathscr{E}$ in (2.5), then $\hat{\gamma}_{2m}^T = \gamma_{2m}^T$.

Proof. Let us assume that there exists $z \in \mathbb{R}$ and $V \in \mathbb{S}^{\sigma(n,m)}$ satisfying the constraint in (3.26), for some vectors $\alpha^{(1)}, \ldots, \alpha^{(r)} \in \mathbb{R}^{\omega(n,m)}$. Then, by setting $v(x) = x^{\{m\}'} V x^{\{m\}}$ one has for all $x \in \mathbb{R}_0^n$ and for all $i = 1, \ldots, r$,

$$
\begin{aligned}
\left. \frac{\dot{v}(x)}{v(x)} \right|_{p=p^{(i)}} &= \frac{x^{\{m\}'} \text{he}\left(V \hat{A}_i^{\#}\right) x^{\{m\}}}{x^{\{m\}'} V x^{\{m\}}} \\
&= \frac{x^{\{m\}'} \left(\text{he}\left(V \hat{A}_i^{\#}\right) + L(\alpha^{(i)})\right) x^{\{m\}}}{x^{\{m\}'} V x^{\{m\}}} \\
&\leq z.
\end{aligned}
$$

Therefore, by exploiting convexity of $\mathcal{P}$, one can conclude that

$$
\frac{\dot{v}(x)}{v(x)} < z \quad \forall x \neq 0_n \ \forall p \in \mathcal{P} \tag{3.27}
$$

which implies that $\gamma_{2m}^T \leq z$.

Now, let us suppose that $(n, 2m) \in \mathcal{E}$. From (3.25) we have that for any $z > \gamma_{2m}^T$ there exists a form $v(x)$ of degree $2m$ in x such that $v(x)$ is positive definite and (3.27) holds. Then, Theorem 2.4 implies that $v(x)$ and $zv(x) - \dot{v}(x)$ admit positive definite SMR matrices for any $p \in \mathcal{P}$, and therefore the constraint in problem (3.26) admits a feasible solution for any $z > \gamma_{2m}^T$. $\qquad \square$

3.4 Rational Parametric Uncertainty

In this section we address the case of time-varying systems affected by rational parametric uncertainty. Specifically, we consider the continuous-time system described by

$$
\dot{x}(t) = A_{rat}(p(t))x(t) \tag{3.28}
$$

where the uncertain parameter vector $p(t)$ is supposed to satisfy

$$
\begin{cases}
p(t) \in \mathcal{P}, & \forall t \geq 0 \\
\mathcal{P} = \text{co}\{p^{(1)}, \ldots, p^{(r)}\}
\end{cases} \tag{3.29}
$$

for some given vectors $p^{(1)}, \ldots, p^{(r)} \in \mathbb{R}^q$, and the matrix function $A_{rat}(p(t)) \in \mathbb{R}^{n \times n}$ has the form

$$
A_{rat}(p(t)) = \frac{1}{a_2(p(t))} A_1(p(t)) \tag{3.30}
$$

where $A_1(p(t)) \in \mathbb{R}^{n \times n}$ and $a_2(p(t)) \in \mathbb{R}$ are respectively a matrix polynomial and a scalar polynomial in the parameters $p(t)$.

Example 3.3. Let us consider the system

$$\begin{cases} \dot{x}_1(t) = x_2(t) \\ \dot{x}_2(t) = -\dfrac{1+2p^2(t)}{1+p^2(t)}x_1(t) - \dfrac{1}{1+p^2(t)}x_2(t). \end{cases}$$

This system can be expressed as in (3.28) with

$$A_1(p(t)) = \begin{pmatrix} 0 & 1+p^2(t) \\ -1-2p^2(t) & -1 \end{pmatrix}$$
$$a_2(p(t)) = 1+p^2(t).$$

Robustness properties of the system (3.28) can be investigated in several ways. A first possibility consists of computing a convex polytope of matrices $\mathscr{A}$ bounding the set of matrices $A_{rat}(p(t))$ in (3.28), i.e. satisfying

$$\mathscr{A} \supseteq \{A_{rat}(p(t)) : p(t) \in \mathscr{P}\} \tag{3.31}$$

and to employ robustness analysis techniques developed in the previous sections for time-varying polytopic systems. However, since $A_{rat}(p(t))$ depends rationally on $p(t)$, this approach leads in general to conservative results.

In the following, we take a different approach based on an LFR of the system (3.28). More specifically, by exploiting LFRs, the system (3.28) can be equivalently rewritten in the following form.

Definition 3.9 (Time-varying LFR System). Consider the continuous-time system described by

$$\begin{cases} \dot{x}(t) = Ax(t) + By(t) \\ z(t) = Cx(t) + Dy(t) \\ y(t) = E(p(t))z(t) \\ E(p(t)) = \begin{pmatrix} p_1(t)I_{s_1} & & \\ & \ddots & \\ & & p_q(t)I_{s_q} \end{pmatrix} \end{cases} \tag{3.32}$$

where $x(t) \in \mathbb{R}^n$ is the state vector; $y(t), z(t) \in \mathbb{R}^d$ are auxiliary vectors with $d = s_1 + \cdots + s_q$ and nonnegative integers $s_1, \ldots, s_q$; $p(t)$ is an uncertain parameter vector supposed to satisfy (3.29) for some given vectors $p^{(1)}, \ldots, p^{(r)} \in \mathbb{R}^q$; and $A \in \mathbb{R}^{n \times n}$, $B \in \mathbb{R}^{n \times d}$, $C \in \mathbb{R}^{d \times n}$, $D \in \mathbb{R}^{d \times d}$ are given matrices. Then, (3.32) is called *time-varying LFR system*.

For the system (3.32) we define the *LFR degree* as

$$d_{LFR} = \max \{s_1, \ldots, s_q\}. \tag{3.33}$$

We also introduce the polytope of matrices

$$\mathscr{P}_E = \{E(p(t)) \in \mathbb{S}^d : p(t) \in \mathscr{P}\} \tag{3.34}$$

whose vertices are given by

$$E_i = E(p^{(i)}), \quad i = 1, \ldots, r. \tag{3.35}$$

The problem of constructing a Lyapunov function for the system (3.28) is hence cast into the problem of constructing a Lyapunov function for the system (3.32).

Example 3.4. Let us consider the system in Example 3.3. It can be verified that this system can be expressed as in (3.32) with $q = 1$, $s_1 = 2$ and

$$A = \begin{pmatrix} 0 & 1 \\ -1 & -1 \end{pmatrix}, \quad B = \begin{pmatrix} 0 & 0 \\ 1 & 0 \end{pmatrix}, \quad C = \begin{pmatrix} 0 & 0 \\ -1 & 1 \end{pmatrix}, \quad D = \begin{pmatrix} 0 & 1 \\ -1 & 0 \end{pmatrix}.$$

In the sequel it will be assumed that the polytope $\mathscr{P}$ contains the origin, i.e. $0_q \in \mathscr{P}$. This means that there is no loss of generality in considering that the matrix A is Hurwitz. In the following, the dependence on the time t will be omitted for ease of presentation.

3.4.1 SMR for LFR Systems

In order to illustrate the main idea, let us first observe that from (3.32) one can write

$$\begin{aligned} \frac{dx^{\{m\}}}{dt} &= \frac{\partial x^{\{m\}}}{\partial x}(Ax + BE(p)z) \\ &= A^{\#}x^{\{m\}} + B_1(p)\left(z \otimes x^{\{m-1\}}\right) \end{aligned} \tag{3.36}$$

where $A^{\#}$ is the extended matrix of A defined by (3.8) and $B_1(p)$ is a suitable matrix independent of x, z. Now, let us consider the form $v(x) = x^{\{m\}'}Vx^{\{m\}}$ for some $V \in \mathbb{S}^{\sigma(n,m)}$. It turns out that

$$\dot{v}(x) = 2\left(x^{\{m\}'}VA^{\#}x^{\{m\}} + x^{\{m\}'}VB_1(p)\left(z \otimes x^{\{m-1\}}\right)\right).$$

This means that $\dot{v}(x)$ can be written as a quadratic form in the vector $f(x,z) \in \mathbb{R}^{\sigma_{LFR}(n,d,m)}$

$$f(x,z) = \begin{pmatrix} x^{\{m\}} \\ z \otimes x^{\{m-1\}} \end{pmatrix}$$

where

$$\sigma_{LFR}(n,d,m) = \sigma(n,m) + d\sigma(n,m-1). \tag{3.37}$$

Indeed,

$$\dot{v}(x) = f(x,z)'\begin{pmatrix} \mathrm{he}\left(VA^{\#}\right) & VB_1(p) \\ \star & 0 \end{pmatrix}f(x,z). \tag{3.38}$$

The representation (3.38) is a sort of SMR with respect to the vector $f(x,z)$. Similarly to the SMR introduced in Chapter 1, the matrix representing $\dot{v}(x)$ in (3.38) is not unique, and the set of such matrices is an affine space. In order to parametrize this set, let us define

$$\mathscr{L}_{n,d,m}^{LFR} = \left\{ L \in \mathbb{S}^{\sigma_{LFR}(n,d,m)} : f(x,z)'Lf(x,z) = 0 \; \forall x \in \mathbb{R}^n \; \forall z \in \mathbb{R}^d \right\}. \quad (3.39)$$

The following result characterizes the set $\mathscr{L}_{n,d,m}^{LFR}$.

Theorem 3.11. *The set $\mathscr{L}_{n,d,m}^{LFR}$ is a linear space of dimension*

$$\begin{aligned}
\omega_{LFR}(n,d,m) &= \frac{1}{2}\sigma_{LFR}(n,d,m)(\sigma_{LFR}(n,d,m)+1) - \sigma(n,2m) \\
&\quad -d\sigma(n,2m-1) - \frac{1}{2}d(d+1)\sigma(n,2m-2).
\end{aligned} \quad (3.40)$$

Proof. The first right hand side term in (3.40) is the number of distinct components of a symmetric matrix with size $\sigma_{LFR}(n,d,m) \times \sigma_{LFR}(n,d,m)$, while the absolute value of the sum of the three negative terms is the number of distinct monomials in the form $f(x,z)'Lf(x,z)$ (which have degree $2m$ in the pair (x,z)). Since the coefficients of the form depend linearly on the components of L, we hence obtain a system of linear equations, which can be shown to be independent, by following the same reasoning as in the proof of Theorem 1.2. Therefore, it follows that the dimension of $\mathscr{L}_{n,d,m}^{LFR}$ is given by (3.40). $\qquad\square$

According to Theorem 3.11, $\mathscr{L}_{n,d,m}^{LFR}$ admits a linear parametrization $L(\alpha)$, where $\alpha \in \mathbb{R}^{\omega_{LFR}(n,d,m)}$ is a free vector. This parametrization can be computed by applying standard algorithms of linear algebra, similar to that reported in Appendix B for scalar forms. As it will be shown in the following, such a parametrization plays a key role in deriving sufficient conditions for establishing robust stability of the system (3.32) because it allows one to take into account the degrees of freedom in the representation (3.38). In particular, let us observe that

$$\begin{aligned}
\sigma_{LFR}(n,d,1) &= n+d \\
\omega_{LFR}(n,d,1) &= 0
\end{aligned}$$

i.e. in the case of quadratic Lyapunov functions (which correspond to the choice $m = 1$) the quantity $\sigma_{LFR}(n,d,m)$ is the sum of the number of states n and the size d of the uncertainty block $E(p)$, while $L(\alpha)$ is identically zero. Tables 3.2 and 3.3 show $\sigma_{LFR}(n,d,m)$ and $\omega_{LFR}(n,d,m)$ for some values of n,d,m.

In order to work out the conditions to enforce negative definiteness of $\dot{v}(x)$ in (3.38), we need to introduce the following matrices.

Definition 3.10 (Extended Matrices for LFR System). Let $B^{\#} \in \mathbb{R}^{\sigma(n,m) \times d\sigma(n,m-1)}$, $C^{\#} \in \mathbb{R}^{d\sigma(n,m-1) \times \sigma(n,m)}$ and $D^{\#} \in \mathbb{R}^{d\sigma(n,m-1) \times d\sigma(n,m-1)}$ be the matrices satisfying the relations

Table 3.2 Quantity $\sigma_{LFR}(n,d,m)$ for some values of n,d,m: (a) $m=2$; (b) $m=3$

	(a)					(b)			
	$d=1$	$d=2$	$d=3$	$d=4$		$d=1$	$d=2$	$d=3$	$d=4$
$n=1$	2	3	4	5	$n=1$	2	3	4	5
$n=2$	5	7	9	11	$n=2$	7	10	13	16
$n=3$	9	12	15	18	$n=3$	16	22	28	34
$n=4$	14	18	22	26	$n=4$	30	40	50	60

Table 3.3 Quantity $\omega_{LFR}(n,d,m)$ for some values of n,d,m: (a) $m=2$; (b) $m=3$

	(a)					(b)			
	$d=1$	$d=2$	$d=3$	$d=4$		$d=1$	$d=2$	$d=3$	$d=4$
$n=1$	0	0	0	0	$n=1$	0	0	0	0
$n=2$	3	6	10	15	$n=2$	10	21	36	55
$n=3$	14	25	39	56	$n=3$	72	138	225	333
$n=4$	40	66	98	136	$n=4$	290	519	813	1172

$$\frac{\partial x^{\{m\}}}{\partial x} Bz = B^{\#}\left(z \otimes x^{\{m-1\}}\right)$$
$$Cx \otimes x^{\{m-1\}} = C^{\#} x^{\{m\}} \qquad (3.41)$$
$$Dz \otimes x^{\{m-1\}} = D^{\#}\left(z \otimes x^{\{m-1\}}\right).$$

for all $x \in \mathbb{R}^n$ and for all $z \in \mathbb{R}^d$. Then, $B^{\#}, C^{\#}, D^{\#}$ are called *extended matrices for LFR systems* of the matrices B, C, D respectively.

The next result provides explicit formulae for the extended matrices $B^{\#}, C^{\#}, D^{\#}$ analogously to the one given in (3.10) for the extended matrix $A^{\#}$ defined by (3.8).

Theorem 3.12. *The matrices* $B^{\#}, C^{\#}, D^{\#}$ *in* (3.41) *are given by*

$$B^{\#} = \left(K'_m K_m\right)^{-1} K'_m \left(\sum_{i=0}^{m-1} (I_{n^j} \otimes B) F_i \otimes I_{n^{m-1-i}}\right) (I_d \otimes K_{m-1}) \qquad (3.42)$$

$$C^{\#} = \left(C \otimes \left(K'_{m-1} K_{m-1}\right)^{-1} K'_{m-1}\right) K_m \qquad (3.43)$$

$$D^{\#} = D \otimes I_{\sigma(n,m-1)} \qquad (3.44)$$

where K_m *and* K_{m-1} *are defined by* (3.9)*, and* $F_i \in \mathbb{R}^{dn^i \times dn^i}$ *is given by*

$$F_i = \left(I_d \otimes e_1^{(n^i)}, \ldots, I_d \otimes e_{n^i}^{(n^i)}\right)'$$

being $e_j^{(k)}$ the j-th column of I_k.

Proof. Let us consider (3.42). From (3.41) and (3.9) it follows that

$$
K_m B^\# \left(p \otimes x^{\{m-1\}} \right) = \frac{\partial x^{[m]}}{\partial x} Bz
$$
$$
= \left(\sum_{i=0}^{m-1} x^{[i]} \otimes I_n \otimes x^{[m-1-i]} \right) Bz.
$$

For any $i = 0, \ldots, m-1$ it turns out that

$$
\left(x^{[i]} \otimes I_n \otimes x^{[m-1-i]} \right) Bz = x^{[i]} \otimes Bz \otimes x^{[m-1-i]}
$$
$$
= (I_{n^i} \otimes B \otimes I_{n^{m-1-i}}) \left(x^{[i]} \otimes z \otimes x^{[m-1-i]} \right)
$$
$$
= (I_{n^i} \otimes B \otimes I_{n^{m-1-i}}) \left(F_i \left(z \otimes x^{[i]} \right) \otimes x^{[m-1-i]} \right)
$$
$$
= (I_{n^i} \otimes B \otimes I_{n^{m-1-i}}) (F_i \otimes I_{n^{m-1-i}}) \left(z \otimes x^{[m-1]} \right)
$$
$$
= ((I_{n^i} \otimes B) F_i \otimes I_{n^{m-1-i}}) \left(z \otimes x^{[m-1]} \right).
$$

Then, (3.42) follows by observing that

$$
z \otimes x^{[m-1]} = (I_d \otimes K_{m-1}) \left(z \otimes x^{\{m-1\}} \right).
$$

Let us consider (3.43). From (3.41), (3.9) and the equation above, it follows that

$$
(I_d \otimes K_{m-1}) C^\# x^{\{m\}} = Cx \otimes x^{[m-1]}
$$
$$
= (C \otimes I_{n^{m-1}}) x^{[m]}
$$
$$
= (C \otimes I_{n^{m-1}}) K_m x^{\{m\}}.
$$

Therefore,

$$
C^\# = \left((I_d \otimes K'_{m-1})(I_d \otimes K_{m-1}) \right)^{-1} (I_d \otimes K'_{m-1})(C \otimes I_{n^{m-1}}) K_m
$$
$$
= \left(I_d \otimes (K'_{m-1} K_{m-1})^{-1} \right) (I_d \otimes K'_{m-1})(C \otimes I_{n^{m-1}}) K_m
$$
$$
= \left(C \otimes (K'_{m-1} K_{m-1})^{-1} K'_{m-1} \right) K_m.
$$

Finally, (3.44) follows from (3.41) and properties of the Kronecker's product. $\qquad \square$

3.4.2   Conditions for Robust Stability

For the system (3.32) we define the property of robust stability analogously to what has been done in Definition 3.2 for the time-varying polytopic system (3.1)–(3.4). The next theorem provides a sufficient condition for establishing robust stability in terms of an infinite number of LMIs.

Theorem 3.13. *Let $L(\cdot)$ be a linear parametrization of $\mathscr{L}_{n,d,m}^{LFR}$ in (3.39). Let us suppose that there exists $V \in \mathbb{S}^{\sigma(n,m)}$ such that, for any $E \in \mathscr{P}_E$, there exist $G_E \in \mathbb{R}^{\sigma(n,m) \times d\sigma(n,m-1)}$, $H_E \in \mathbb{R}^{d\sigma(n,m-1) \times d\sigma(n,m-1)}$ and $\alpha_E \in \mathbb{R}^{\omega_{LFR}(n,d,m)}$ satisfying*

$$\begin{cases} 0 < V \\ 0 > R(V, G_E, H_E, E) + L(\alpha_E) \end{cases} \tag{3.45}$$

where

$$R(V, G_E, H_E, E) = U(V, E) + V(G_E, H_E, E)$$
$$U(V, E) = \begin{pmatrix} \mathrm{he}(VA^{\#}) & V(BE)^{\#} \\ \star & 0_{d\sigma(n,m-1) \times d\sigma(n,m-1)} \end{pmatrix}$$
$$V(G_E, H_E, E) = \begin{pmatrix} \mathrm{he}(G_E C^{\#}) & G_E(DE)^{\#} - G_E + (H_E C^{\#})' \\ \star & \mathrm{he}(H_E(DE)^{\#} - H_E) \end{pmatrix}$$

and $(BE)^{\#}$ and $(DE)^{\#}$ denote the extended matrices of BE and DE, respectively, defined according to (3.42) and (3.44). Then, the system (3.32) is robustly stable.

Proof. Let us suppose that (3.45) holds for some V and for some G_E, H_E, α_E which depend on E. Let us define $v(x) = x^{\{m\}'} V x^{\{m\}}$. Since $V > 0$ we have that $v(x)$ is positive definite. Then, from (3.32), (3.36) and (3.41) one has

$$\dot{v}(x) = f(x,z)' U(V, E) f(x,z).$$

By (3.32), one has $z = Cx + DE(p)z$. Then, from (3.41) it follows that for any G_E and H_E one can write

$$x^{\{m\}'} G_E \left(z \otimes x^{\{m-1\}} \right) = x^{\{m\}'} G_E \left(C^{\#} x^{\{m\}} + (DE)^{\#} \left(z \otimes x^{\{m-1\}} \right) \right)$$
$$\left(z \otimes x^{\{m-1\}} \right)' H_E \left(z \otimes x^{\{m-1\}} \right) = \left(z \otimes x^{\{m-1\}} \right)' H_E \left(C^{\#} x^{\{m\}} + (DE)^{\#} \left(z \otimes x^{\{m-1\}} \right) \right)$$

or, equivalently,

$$f(x,z)' V(G_E, H_E, E) f(x,z) = 0.$$

Since $f(x,z)' L(\alpha_E) f(x,z) = 0$ for any α_E, we can also write that

$$\dot{v}(x) = f(x,z)' \left(U(V, E) + V(G_E, H_E, E) + L(\alpha_E) \right) f(x,z)$$

or, equivalently,

$$\dot{v}(x) = f(x,z)' \left(R(V, G_E, H_E, E) + L(\alpha_E) \right) f(x,z).$$

Hence, for all $E \in \mathscr{P}_E$ we have that $\dot{v}(x)$ is negative definite. Therefore, the system (3.32) is robustly stable. □

Theorem 3.13 requires to solve an infinite number of LMIs because the matrices G_E and H_E are allowed to change for different $E \in \mathscr{P}_E$. Sufficient conditions for robust stability based on a finite number of LMIs can be derived via a suitable choice of G_E and H_E. The case of constant matrices is considered next.

Theorem 3.14. *Let $L(\cdot)$ be a linear parametrization of $\mathscr{L}_{n,d,m}^{LFR}$ in (3.39). Let us suppose that there exist $V \in \mathbb{S}^{\sigma(n,m)}$, $G \in \mathbb{R}^{\sigma(n,m) \times d\sigma(n,m-1)}$, $H \in \mathbb{R}^{d\sigma(n,m-1) \times d\sigma(n,m-1)}$ and $\alpha^{(i)} \in \mathbb{R}^{\omega_{LFR}(n,d,m)}$, $i = 1, \ldots, r$, such that*

$$\begin{cases} 0 < V \\ 0 > R(V,G,H,E_i) + L(\alpha^{(i)}), \quad i = 1, \ldots, r. \end{cases} \tag{3.46}$$

Then, the system (3.32) is robustly stable.

Proof. Let us consider any $E \in \mathscr{P}_E$. Since $\mathscr{P}_E$ is the convex hull of the matrices $E_1, \ldots, E_r$ in (3.35), we have that E can be written as

$$E = \sum_{i=1}^{r} c_i(E) E_i$$

where $c_1(E), \ldots, c_r(E) \in \mathbb{R}$ are such that

$$\begin{cases} \sum_{i=1}^{r} c_i(E) E_i = E \\ \sum_{i=1}^{r} c_i(E) = 1 \\ c_i(E) \geq 0 \quad \forall i = 1, \ldots, r. \end{cases} \tag{3.47}$$

Let us define

$$\alpha_E = \sum_{i=1}^{r} c_i(E) \alpha^{(i)}.$$

Following the same reasoning as in the proof of Theorem 3.13, we choose $v(x)) = x^{\{m\}'} V x^{\{m\}}$. Then, we have

$$\begin{aligned} \dot{v}(x) &= f(x,z)' R(V,G,H,E) f(x,z) \\ &= f(x,z)' \left(R(V,G,H,E) + L(\alpha_E) \right) f(x,z) \\ &= \sum_{i=1}^{r} c_i(E) g_i(x) \end{aligned}$$

where

$$g_i(x) = f(x,z)' \left(R(V,G,H,E_i) + L(\alpha^{(i)}) \right) f(x,z).$$

Since $g_i(x) < 0$ by (3.46) and $c_i(E)$ satisfies (3.47), it follows that $\dot{v}(x)$ is negative definite. □

In the LFR degree is $d_{LFR} = 1$, a less stringent condition can be obtained as stated next.

Theorem 3.15. *Suppose that $d_{LFR} = 1$, and let $L(\cdot)$ be a linear parametrization of $\mathscr{L}_{n,d,m}^{LFR}$ in (3.39). Let us suppose that there exist $V \in \mathbb{S}^{\sigma(n,m)}$, $G_i \in \mathbb{R}^{\sigma(n,m) \times d\sigma(n,m-1)}$, $H_i \in \mathbb{R}^{d\sigma(n,m-1) \times d\sigma(n,m-1)}$ and $\alpha^{(i)} \in \mathbb{R}^{\omega_{LFR}(n,d,m)}$, $i = 1,\dots,r$, such that*

$$\begin{cases} 0 < V \\ 0 > R(V, G_i, H_i, E_i) + L(\alpha^{(i)}), & i = 1,\dots r. \end{cases} \tag{3.48}$$

Then, the system (3.32) is robustly stable.

Proof. Let us define $v(x) = x^{\{m\}'} V x^{\{m\}}$. Since the system (3.32) can be rewritten as $\dot{x} = \bar{A}(E(p))x$ where

$$\bar{A}(E) = A + BE(I - DE)^{-1}C,$$

it follows that

$$\dot{v}(x) = x^{\{m\}'} \, \mathrm{he}\left(V\bar{A}(E)^{\#}\right) x^{\{m\}}$$

where $\bar{A}(E)^{\#}$ is the extended matrix of $\bar{A}(E)$, defined according to (3.8). From the proof of Theorem 3.13, (3.48) implies

$$\dot{v}(x)|_{E=E_i} < 0 \quad \forall x \neq 0_n \,\, \forall i = 1,\dots r.$$

One also has

$$\dot{v}(x)|_{E=E_i} = x^{\{m\}'} \, \mathrm{he}\left(V\bar{A}(E_i)^{\#}\right) x^{\{m\}}, \quad i = 1,\dots r$$

where $\bar{A}(E_i)^{\#}$ is the extended matrix of $\bar{A}(E_i)$, defined according to (3.8). Since $d_{LFR} = 1$, it turns out that (see e.g. [16])

$$\mathrm{co}\left\{\bar{A}(E_i), \, i = 1,\dots,r\right\} = \left\{\bar{A}(E), \, E \in \mathrm{co}\{E_1,\dots,E_r\}\right\}.$$

This implies that $\bar{A}(E)^{\#}$ can be expressed as a convex combination of $\bar{A}(E_1)^{\#},\dots,$ $\bar{A}(E_r)^{\#}$ for any admissible E. Moreover, since $\dot{v}(x)$ is linear in $\bar{A}(E)^{\#}$, it follows that

$$\dot{v}(x) < 0 \quad \forall x \neq 0_n \,\, \forall E \in \mathrm{co}\{E_1,\dots,E_r\}$$

i.e. the system (3.32) is robustly stable. □

The LMI feasibility test (3.46) involves a number of scalar variables equal to

$$\sigma(n,m)\left(\frac{\sigma(n,m)+1}{2} + d\sigma(n,m-1)\right) + d^2\sigma^2(n,m-1) + r\omega_{LFR}(n,d,m)$$

while this number for the LMI feasibility test (3.48) is given by

$$\sigma(n,m)\left(\frac{\sigma(n,m)+1}{2}+dr\sigma(n,m-1)\right)+d^2r\sigma^2(n,m-1)+r\omega_{LFR}(n,d,m).$$

The previous results extend to the HPLFs setting the results obtained in [144] by using quadratic Lyapunov functions, which correspond to the case $m=1$. The next theorem clarifies that HPLFs are guaranteed to provide less conservative results than quadratic Lyapunov functions.

Theorem 3.16. *If condition* (3.45) *of Theorem 3.13 holds for $m=1$, then it holds also for any integer $m>1$.*

Proof. Let $\bar{V},\bar{G}_E,\bar{H}_E$ be the matrices V,G_E,H_E satisfying (3.45) for $m=1$. First of all, let us observe that $\mathscr{L}_{n,d,1}^{LFR}$ is empty, i.e. $L(\alpha_E)=0$ in (3.45). Therefore, we can write

$$R(\bar{V},\bar{G}_E,\bar{H}_E,E)<0.$$

Let us consider $m=2$ for simplicity, and let us select

$$V=K_2'\bar{V}^{[2]}K_2$$

where K_2 is defined by (3.9). Since K_2 has full column rank, and, by assumption, $\bar{V}>0$, it follows that $V>0$. Let us select

$$\begin{cases} G_E=2K_2(\bar{G}_E\otimes\bar{V}) \\ H_E=2\bar{H}_E\otimes\bar{V}. \end{cases}$$

Then, from Theorem 3.13 it turns out that

$$\begin{cases} R(V,G_E,H_E,E)=2X \\ X=\begin{pmatrix} K_2 \\ & I_{dn} \end{pmatrix}'(R(\bar{V},\bar{G}_E,\bar{H}_E,E)\otimes\bar{V})\begin{pmatrix} K_2 \\ & I_{dn} \end{pmatrix}. \end{cases}$$

Therefore, it follows that

$$R(V,G_E,H_E,E)<0$$

i.e. the sufficient condition of Theorem 3.13 is satisfied also for $m=2$. The same reasoning can be repeated for any $m>2$. $\square$

The proof of Theorem 3.16 clarifies that the same reasoning can be applied for any choice of matrices G_E and H_E, and therefore a similar result holds also for Theorems 3.14 and 3.15.

3.4.3 Robust Stability Margin for LFR Systems

The aim of this section is to show how Theorem 3.14 and Theorem 3.15 can be exploited for computing a lower bound of the polytopic stability margin, for system (3.32). In particular, we denote by *LFR stability margin* for such a system, the quantity

$$\gamma^{PR} = \sup \ \{\gamma \in \mathbb{R} : (3.32) \text{ is robustly stable for all } p(t) \in \gamma \mathscr{P}\}.$$

By using HPLFs of degree $2m$, we define the lower bound to γ^{PR}

$$\gamma_{2m}^{PR} = \sup \ \{\gamma \in \mathbb{R} : \exists v \in \Xi_{n,2m} \text{ HPLF for (3.32), with } p(t) \in \gamma \mathscr{P}\}$$

which we refer to as *$2m$-HPLF LFR stability margin*. The following result shows how a lower bound to γ_{2m}^{PR} can be computed by solving a GEVP.

Theorem 3.17. *Let $m \geq 1$ be an integer, and $L(\cdot)$ be a linear parametrization of $\mathscr{L}_{n,d,m}^{LFR}$ in (3.39). Let us define*

$$\hat{\gamma}_{2m}^{PR} = \frac{1}{z^*}$$

where z^* is the solution of the GEVP

$$
z^* = \inf_{\substack{z \in \mathbb{R}, \ V \in \mathbb{S}^{\sigma(n,m)}, \\ G \in \mathbb{R}^{\sigma(n,m) \times d\sigma(n,m-1)}, \\ H \in \mathbb{R}^{d\sigma(n,m-1) \times d\sigma(n,m-1)}, \\ \alpha^{(0)}, \ldots, \alpha^{(r)} \in \mathbb{R}^{\omega_{LFR}(n,d,m)}}} z \qquad (3.49)
$$

$$
\text{s.t.} \ \begin{cases} 0 < V \\ 0 < -R(V,G,H,0) - L(\alpha^{(0)}) \\ 0 < z\left(-R(V,G,H,0) - L(\alpha^{(0)})\right) - R(V,G,H,E_i) \\ \quad + R(V,G,H,0) - L(\alpha^{(i)}), \quad i = 1,\ldots,r. \end{cases}
$$

Then, $\hat{\gamma}_{2m}^{PR} \leq \gamma_{2m}^{PR}$.

Proof. Let us consider any $\gamma \leq \hat{\gamma}_{2m}^{PR}$ and define $\hat{\alpha}^{(i)} = \gamma \alpha^{(i)} + \alpha^{(0)}$, for $i = 1,\ldots,r$. It follows that

$$
R(V,G,H,\gamma E_i) + L(\hat{\alpha}^{(i)})
$$
$$
= \gamma\left(R(V,G,H,E_i) - R(V,G,H,0) + L(\alpha^{(i)})\right) + R(V,G,H,0) + L(\alpha^{(0)})
$$
$$
< 0
$$

i.e. (3.46) holds with E_i and $\alpha^{(i)}$ replaced by γE_i and $\hat{\alpha}^{(i)}$ respectively. Hence, the result follows from Theorem 3.14. $\qquad \square$

It is worth observing that the constraint $-R(V,G,H,0) - L(\alpha^{(0)}) > 0$ in (3.49), which is required in order to ensure that (3.49) is a GEVP, is not restrictive because $\mathscr{P}_E$ includes the origin.

In the case when the LFR degree is $d_{LFR} = 1$, a tighter lower bound with respect to that provided by Theorem 3.17 can be obtained as follows.

Theorem 3.18. *Let $m \geq 1$ be an integer, and $L(\cdot)$ be a linear parametrization of $\mathscr{L}^{LFR}_{n,d,m}$ in (3.39). Let us define*

$$\hat{\gamma}^{PRO}_{2m} = \frac{1}{z^*}$$

where z^* is the solution of the GEVP

$$
z^* = \inf_{\substack{z \in \mathbb{R},\ V \in \mathbb{S}^{\sigma(n,m)}, \\ G_i \in \mathbb{R}^{\sigma(n,m) \times d\sigma(n,m-1)}, \\ H_i \in \mathbb{R}^{d\sigma(n,m-1) \times d\sigma(n,m-1)}, \\ \alpha^{(i)}, \hat{\alpha}^{(i)} \in \mathbb{R}^{\omega_{LFR}(n,d,m)},\ i=1,\ldots,r}} z
$$

$$
s.t. \quad
\begin{cases}
0 < V \\
0 < -R(V,G_i,H_i,0) - L(\alpha^{(i)}), \quad i=1,\ldots,r \\
0 < z\left(-R(V,G_i,H_i,0) - L(\alpha^{(i)})\right) - R(V,G_i,H_i,E_i) \\
\quad + R(V,G_i,H_i,0) - L(\hat{\alpha}^{(i)}), \quad i=1,\ldots,r.
\end{cases}
\tag{3.50}
$$

Then, $\hat{\gamma}^{PRO}_{2m} \leq \gamma^{PR}_{2m}$.

Proof. The proof follows from Theorem 3.15 and arguments analogous to the proof of Theorem 3.17.   □

The GEVP (3.49) involves a number of scalar variables equal to

$$\sigma(n,m)\left(\frac{\sigma(n,m)+1}{2} + d\sigma(n,m-1)\right) + d^2\sigma^2(n,m-1) + (r+1)\omega_{LFR}(n,d,m)$$

while this number for the GEVP (3.50) is given by

$$\sigma(n,m)\left(\frac{\sigma(n,m)+1}{2} + dr\sigma(n,m-1)\right) + d^2 r\sigma^2(n,m-1) + 2r\omega_{LFR}(n,d,m).$$

3.5 Examples

In this section we present some examples in which the proposed conditions for robustness analysis of uncertain systems with time-varying parametric uncertainty are applied.

3.5.1 Example HPLF-1

Let us assume we want to compute the stability margin κ_{2m}^∞ defined in (3.21) for the segment of matrices (3.20), with

$$A_0 = \begin{pmatrix} 0 & 1 \\ -2 & -1 \end{pmatrix}, \quad A_1 = \begin{pmatrix} 0 & 0 \\ -1 & 0 \end{pmatrix}.$$

Since in this example we have $n = 2$, the lower bound provided by (3.22) is tight, i.e. $\hat{\kappa}_{2m}^\infty = \kappa_{2m}^\infty$, according to Corollary 3.1.

Table 3.4 shows values of κ_{2m}^∞ obtained by solving the GEVP (3.22) for different values of m and the corresponding number of scalar variables.

Table 3.4 Example HPLF-1: stability margin κ_{2m}^∞ and number of scalar variables in (3.22)

m	1	2	3	4	5	6	7	8	9	10
κ_{2m}^∞	3.8284	5.7393	6.2135	6.3982	6.6469	6.6580	6.7800	6.7961	6.8412	6.8649
# variables	4	9	17	28	42	59	79	102	128	157

In order to illustrate the main features of HPLFs, let us consider in more detail the case $m = 3$. The power vector $x^{\{3\}}$ and the matrix $L(\alpha)$ are chosen as

$$x^{\{3\}} = \begin{pmatrix} x_1^3 \\ x_1^2 x_2 \\ x_1 x_2^2 \\ x_2^3 \end{pmatrix}, \quad L(\alpha) = \begin{pmatrix} 0 & 0 & -\alpha_1 & -\alpha_2 \\ \star & 2\alpha_1 & \alpha_2 & -\alpha_3 \\ \star & \star & 2\alpha_3 & 0 \\ \star & \star & \star & 0 \end{pmatrix}.$$

Hence, the extended matrices of A_0 and A_1 are equal to

$$A_0^\# = \begin{pmatrix} 0 & 3 & 0 & 0 \\ -2 & -1 & 2 & 0 \\ 0 & -4 & -2 & 1 \\ 0 & 0 & -6 & -3 \end{pmatrix}, \quad A_1^\# = \begin{pmatrix} 0 & 0 & 0 & 0 \\ -1 & 0 & 0 & 0 \\ 0 & -2 & 0 & 0 \\ 0 & 0 & -3 & 0 \end{pmatrix}.$$

The matrix V obtained by solving (3.22) for $m = 3$ turns out to be

$$V = \begin{pmatrix} 1.0000 & 0.1524 & -0.1965 & -0.0174 \\ \star & 1.2187 & 0.2086 & -0.0280 \\ \star & \star & 0.2938 & 0.0231 \\ \star & \star & \star & 0.0099 \end{pmatrix}$$

corresponding to the HPLF

$$v(x) = x_1^6 + 0.3048x_1^5x_2 + 0.8258x_1^4x_2^2 + 0.3825x_1^3x_2^3 + 0.2378x_1^2x_2^4$$
$$+ 0.0463x_1x_2^5 + 0.0099x_2^6. \tag{3.51}$$

Figures 3.1a and 3.1b show the trajectories of the system for, respectively, $p(t) \equiv 0$ and $p(t) \equiv 6.2135$. Figures 3.2a and 3.2b show level curves of the quadratic Lyapunov function $v(x) = x_1^2 + 0.2555x_1x_2 + 0.2555x_2^2$, guaranteeing robust stability for $\kappa < 3.8284$, and of the HPLF in (3.51) achieving $\kappa_6^\infty = 6.2135$.

It is worth recalling that this example has been tackled in the literature by adopting a number of techniques employing different classes of Lyapunov functions. In [148], it has been observed that a quadratic Lyapunov function exists only for $\kappa < 3.82$, and that this bound can be improved to $\kappa < 5.73$ by using a quartic Lyapunov function. In [7], a polyhedral Lyapunov function has been constructed, guaranteeing asymptotic stability for $\kappa = 6$. In [146], a piecewise quadratic Lyapunov function achieving stability for $\kappa = 6.2$, has been obtained via a sequence of LMI problems and a grid search over two free parameters. As it can be seen from Table 3.4, these bounds are improved by all the HPLFs of degree greater than 2.

Lastly, it is useful observing that the exact value of κ^∞ for this example has been found in [92] which proposes a necessary and sufficient conditions for robust stability in the case of second-order systems, and is equal to $\kappa^\infty = 6.8951$.

3.5.2 Example HPLF-2

In this example, we want to compute the stability margin κ_{2m}^∞ for the segment of matrices (3.20), with

$$A_0 = \begin{pmatrix} 0 & 1 & 0 \\ 0 & 0 & 1 \\ -1 & -2 & -4 \end{pmatrix}, \quad A_1 = \begin{pmatrix} -2 & 0 & -1 \\ 1 & -10 & 3 \\ 3 & -4 & 2 \end{pmatrix}. \tag{3.52}$$

Quadratic stability is guaranteed only for $\kappa \leq 1.9042$. Let us consider an HPLF of degree 4. Being $n = 3$ and $m = 2$ it follows that $d = 6$ and $\omega(n,m) = 6$. Then, the power vector $x^{\{2\}}$ and the matrix $L(\alpha)$ are given by

$$x^{\{2\}} = \begin{pmatrix} x_1^2 \\ x_1x_2 \\ x_1x_3 \\ x_2^2 \\ x_2x_3 \\ x_3^2 \end{pmatrix}, \quad L(\alpha) = \begin{pmatrix} 0 & 0 & 0 & -\alpha_1 & -\alpha_2 & -\alpha_3 \\ \star & 2\alpha_1 & \alpha_2 & 0 & -\alpha_4 & -\alpha_5 \\ \star & \star & 2\alpha_3 & \alpha_4 & \alpha_5 & 0 \\ \star & \star & \star & 0 & 0 & -\alpha_6 \\ \star & \star & \star & \star & 2\alpha_6 & 0 \\ \star & \star & \star & \star & \star & 0 \end{pmatrix}.$$

The GEVP (3.22) returns the lower bound $\hat{\kappa}_4^\infty = 75.1071$ and the corresponding matrix V given by

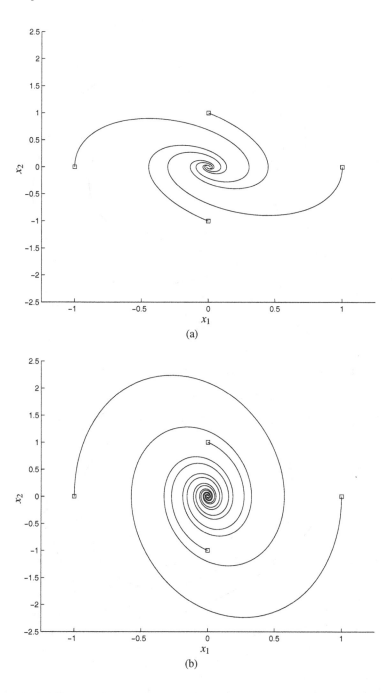

Fig. 3.1 Example HPLF-1: (a) trajectories for $p(t) \equiv 0$; (b) trajectories for $p(t) \equiv 6.2135$

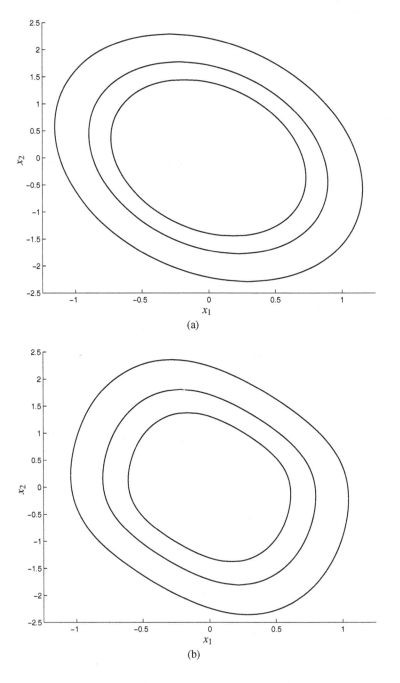

Fig. 3.2 Example HPLF-1: (a) level curves of the quadratic Lyapunov function; (b) level curves of the HPLF in (3.51)

$$V = \begin{pmatrix} 1.0000 & 0.2119 & 0.8931 & -0.2173 & -0.2577 & 0.1695 \\ \star & 3.9362 & -0.7602 & 1.4424 & 2.3656 & -0.7126 \\ \star & \star & 2.3577 & -0.5290 & -1.4494 & 1.0428 \\ \star & \star & \star & 1.6931 & 1.0471 & -0.4101 \\ \star & \star & \star & \star & 3.5386 & -1.1038 \\ \star & \star & \star & \star & \star & 0.7053 \end{pmatrix}.$$

Moreover, since $(n, 2m) = (3, 4) \in \mathscr{E}$, from Corollary 3.1 we have that $\hat{\kappa}_4^\infty = \kappa_4^\infty$.

Using the approach proposed in [148], one finds that the maximum κ for which robust stability is guaranteed is equal to 17.8347. The remarkable difference with respect to our approach is due to the fact that the parametrization of forms adopted in [148] is not complete. Specifically, it is easy to see that in [148] only the parameters $\alpha_1, \alpha_3, \alpha_6$ in the matrix $L(\alpha)$ are considered.

It is worth remarking that in [69] it has been observed that duality can be fruit-fully exploited when addressing robust stability of extended systems. In particular, if one reformulates the robust stability problem in this example by considering the extended matrices of A_0' and A_1', instead of those of A_0 and A_1 in (3.52), the resulting stability margin for an HPLF of degree 4 turns out to be $\hat{\kappa}_4^\infty = +\infty$.

3.5.3 Example HPLF-3

Let us consider the differential equation

$$\dddot{\xi}(t) + \dot{\xi}(t) + k(t)\xi(t) = 0, \quad k(t) \in [0, \kappa]$$

and assume that we want to compute the maximum κ such that the solution remains bounded. By following the reasoning proposed in [18], one finds that the maximum value of κ satisfies

$$\begin{cases} 0 = \sqrt{\kappa}e^a - 1 \\ a = -\dfrac{1}{\sqrt{4\kappa - 1}}\left(\pi - \arctan(\sqrt{4\kappa - 1})\right) \end{cases}$$

and hence is equal to $\kappa^\infty = 3.0448$.

Clearly, the problem can be easily tackled in the framework of HPLF by solving (3.21) with

$$A_0 = \begin{pmatrix} 0 & 1 \\ -\varepsilon & -1 \end{pmatrix}, \quad A_1 = \begin{pmatrix} 0 & 0 \\ -1 & 0 \end{pmatrix}$$

where the scalar $\varepsilon > 0$ is introduced in order to guarantee that A_0 is Hurwitz, which is a necessary condition for ensuring asymptotical stability. Table 3.5 shows values of κ_{2m}^∞ obtained from (3.22) for different values of m with $\varepsilon = 10^{-5}$. Notice that these values exhibit a growth towards κ^∞. On the other hand, Theorem 3.7

guarantees that for any $\kappa < \kappa^\infty$ there exists an HPLF of suitable degree which is also an SOS form; hence it is expected that κ_{2m}^∞ converges to κ^∞ as m approaches infinity.

Table 3.5 Example HPLF-3: stability margin κ_{2m}^∞ (see Table 3.4 for the number of variables)

m	2	3	4	5	6	7	8	9	10	11	12
κ_{2m}^∞	1.5095	2.0121	2.3060	2.4106	2.5158	2.6196	2.6686	2.7042	2.7541	2.7811	2.7898

Another interesting byproduct of the treatment in [18] is that one can calculate the worst-case sequence $k(t)$ for $\kappa = \kappa^\infty$, which prevents the solution $\xi(t)$ from converging to zero, consists of suitable switchings between $k = 0$ and $k = \kappa^\infty$. The resulting trajectory $\xi(t)$ of the system is dashed in Figure 3.3b and represents the limit cycle towards which the level surfaces of the Lyapunov functions are expected to tend. This is confirmed by Figure 3.3a, where the level curves of the obtained HPLFs are depicted for different values of m. Figure 3.3b indeed shows that the HPLF corresponding to $m = 12$ is very close to the limit trajectory predicted by the theory. On the other hand, being the limit trajectory nondifferentiable, it appears clear that the degree of the HPLF can become arbitrarily high as κ approaches κ^∞.

3.5.4 Example HPLF-4

Let us consider system (3.1) with

$$A_0 = \begin{pmatrix} 0 & 1 \\ -1 & -1 \end{pmatrix}, \quad A_1 = \begin{pmatrix} 0 & 0 \\ 1 & 0 \end{pmatrix}.$$

The aim is to compute the $2m$-HPLF ℓ_∞ stability margin γ_{2m}^∞. Notice that $A_0 + A_1$ is not asymptotically stable, which implies that

$$\gamma_{2m}^\infty \leq 1 \quad \forall m \geq 1.$$

By applying Theorem 3.8 for different values of m, one can construct HPLFs of increasing degree, trying to improve the value of the stability margin. Observe that the lower bound provided by (3.19) always coincides with γ_{2m}^∞ because $n = 2$.

For $m = 2$, one obtains $\gamma_4^\infty = 0.9771$, but for $m \geq 3$ the result returned by the GEVP is $\gamma_{2m}^\infty = 1$, which means that it is possible to construct an HPLF of degree 6 (or more) for $|p(t)| \leq \gamma$ and γ arbitrarily close to 1.

In particular, for $m = 3$ the following matrix V has been obtained:

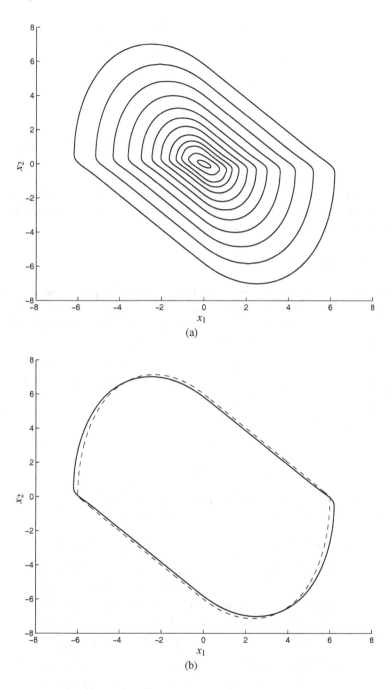

Fig. 3.3 Example HPLF-3: (a) level curves of $v(x)$ for $m = 1, \ldots, 12$, with m increasing from the most inner curve to the most outer; (b) level curve of the HPLF for $m = 12$ (solid line) and limit trajectory (dashed line)

$$V = \begin{pmatrix} 1.0000 & 2.9987 & 1.9974 & 0.2491 \\ \star & 14.1343 & 12.4475 & 2.1782 \\ \star & \star & 14.5114 & 3.5064 \\ \star & \star & \star & 1.3106 \end{pmatrix}$$

corresponding to $\gamma_6^\infty = 1$ and the HPLF

$$v(x) = x_1^6 + 5.9974x_1^5x_2 + 18.1291x_1^4x_2^2 + 25.3932x_1^3x_2^3 + 18.8678x_1^2x_2^4$$
$$+ 7.0128x_1x_2^5 + 1.3106x_2^6.$$

In [119], it has been shown that quadratic Lyapunov functions can achieve $\gamma_2^\infty = \sqrt{3}/2$. In [6], stability for $|p(t)| \leq 0.98$ has been ensured via a polyhedral Lyapunov function whose level sets have 30 vertices.

3.5.5 Example HPLF-5

This example concerns the computation of the best transient performance index γ^T introduced in Definition 3.7. Let us consider the helicopter model originally proposed in [96], and the robust controller designed in [20]. The resulting closed-loop uncertain system is given by $\dot{x}(t) = A(p(t))x(t)$, where

$$A(p) = \begin{pmatrix} -0.1027 & 0.1762 & 0.1995 & -0.3446 \\ -0.7275 - 0.1640p_3 & -1.4280 + 0.2699p_3 & 2.1852 + 0.4511p_3 & 0.7218 + 0.4308p_3 \\ 1.1689 & -0.4446 + p_1 & -3.6757 & -3.0841 + p_2 \\ 0 & 0 & 1 & 0 \end{pmatrix}$$

with

$$|p_1(t)| \leq 0.2192, \quad |p_2(t)| \leq 1.2031, \quad |p_3(t)| \leq 2.0673 \quad \forall t \geq 0.$$

The system can be easily written in form (3.1)–(3.2) with $r = 8$. By solving the GEVP problem (3.26) with $m = 2$, one obtains the HPLF

$$\begin{aligned} v(x) = {} & x_1^4 + 0.2184x_1^3x_2 - 2.8083x_1^3x_3 - 4.0483x_1^3x_4 + 1.1458x_1^2x_2^2 \\ & + 0.1831x_1^2x_2x_3 + 0.3233x_1^2x_2x_4 + 9.7670x_1^2x_3^2 + 16.4758x_1^2x_3x_4 \\ & + 9.7066x_1^2x_4^2 + 0.0283x_1x_2^3 - 2.5961x_1x_2^2x_3 - 3.9673x_1x_2^2x_4 \\ & - 0.1303x_1x_2x_3^2 - 3.6305x_1x_2x_3x_4 - 3.3848x_1x_2x_4^2 - 14.8574x_1x_3^3 \\ & - 37.3189x_1x_3^2x_4 - 31.4426x_1x_3x_4^2 - 11.2239x_1x_4^3 + 0.1064x_2^4 \\ & + 0.1657x_2^3x_3 + 0.2465x_2^3x_4 + 2.1908x_2^2x_3^2 + 5.1952x_2^2x_3x_4 \\ & + 4.2363x_2^2x_4^2 - 0.0017x_2x_3^3 + 3.0142x_2x_3^2x_4 + 6.4605x_2x_3x_4^2 \\ & + 3.4452x_2x_4^3 + 8.8681x_3^4 + 29.0364x_3^3x_4 + 36.5118x_3^2x_4^2 \\ & + 20.6917x_3x_4^3 + 5.8479x_4^4 \end{aligned}$$

achieving $\hat{\gamma}_4^T = -0.8889$.

In [101], the problem of computing the quadratic Lyapunov function achieving the best transient performance, defined as in (3.24), has been addressed. The value obtained was $\hat{\gamma}_2^T = -0.3839$.

3.5.6 Example HPLF-6

Let us consider the LFR system (3.28)–(3.30) with

$$A_{rat}(p(t)) = \begin{pmatrix} 0 & 1 \\ -3 - 4p_2(t) & -\dfrac{2 + p_1(t) + 2p_1(t)p_2(t)}{2 - p_1(t)} \end{pmatrix} \tag{3.53}$$

$$\mathscr{P} = \mathrm{co}\left\{ \begin{pmatrix} 0 \\ 0 \end{pmatrix}, \begin{pmatrix} 1 \\ 3 \end{pmatrix}, \begin{pmatrix} 1 \\ -1 \end{pmatrix} \right\}.$$

This system can be cast into the LFR (3.32) where

$$A = \begin{pmatrix} 0 & 1 \\ -3 & -1 \end{pmatrix}, \quad B = \begin{pmatrix} 0 & 0 \\ 1 & 1 \end{pmatrix}, \quad C = \begin{pmatrix} 0 & -1 \\ -4 & 0 \end{pmatrix}$$

$$D = \begin{pmatrix} 0.5 & 0 \\ 1 & 0 \end{pmatrix}, \quad E(p) = \begin{pmatrix} p_1 & 0 \\ \star & p_2 \end{pmatrix}.$$

It hence follows that $n = 2$, $d = 2$ and $d_{LFR} = 1$. Table 3.6 shows the lower bounds $\hat{\gamma}_{2m}^{PR}$ and $\hat{\gamma}_{2m}^{PRO}$ provided by Theorems 3.17 and 3.18 (the latter exploiting a less stringent condition due to $d_{LFR} = 1$), and the number of scalar variables in the corresponding GEVPs. As we can see, right from $m = 2$ there is a significant improvement with respect to the use of quadratic Lyapunov functions (corresponding to $m = 1$). Figure 3.4 shows, for the same values of m, the level curves of $v(x) = x^{\{m\}'} V x^{\{m\}}$ for the matrices V corresponding to $\hat{\gamma}_{2m}^{PRO}$.

For the sake of comparison, we consider an alternative approach which consists of establishing robust stability of (3.53) over $\hat{\gamma}_{10}^{PRO} \mathscr{P}$ by using the techniques for constructing HPLFs in the case of time-varying polytopic systems. Let us define

$$\mathscr{A}_1 = \{ A_{rat}(p) : p \in \hat{\gamma}_{10}^{PRO} \mathscr{P} \}.$$

A polytope $\hat{\mathscr{A}}_1$ bounding the set $\mathscr{A}_1$ can be obtained as

$$\hat{\mathscr{A}}_1 = \left\{ \begin{pmatrix} 0 & 1 \\ c_1 & c_2 \end{pmatrix} : \begin{pmatrix} c_1 \\ c_2 \end{pmatrix} \in \hat{\mathscr{A}}_2 \right\}$$

where $\hat{\mathscr{A}}_2$ is a polytope bounding the set $\mathscr{A}_2$ defined as

Table 3.6 Example HPLF-6: (a) lower bound $\hat{\gamma}_{2m}^{PR}$ and number of variables in (3.49) ; (b) lower bound $\hat{\gamma}_{2m}^{PRO}$ and number of variables in (3.50)

	(a)			(b)	
m	$\hat{\gamma}_{2m}^{PR}$	# variables	m	$\hat{\gamma}_{2m}^{PRO}$	# variables
1	0.4198	11	1	0.4267	27
2	0.5969	58	2	0.6147	126
3	0.6547	154	3	0.6958	316
4	0.6784	299	4	0.7167	597
5	0.6820	493	5	0.7216	969

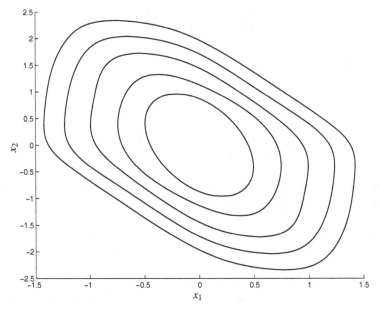

Fig. 3.4 Example HPLF-6: level curves of $v(x)$ corresponding to $\hat{\gamma}_{2m}^{PRO}$ for $m = 1,\ldots,5$, with m increasing from the most inner curve to the most outer

$$\mathscr{A}_2 = \left\{ \begin{pmatrix} c_1 \\ c_2 \end{pmatrix} : c_1 = -3 - 4p_2,\ c_2 = -\frac{2 + p_1 + 2p_1 p_2}{2 - p_1},\ p \in \hat{\gamma}_{10}^{PRO} \mathscr{P} \right\}.$$

It has been found that robust stability of the time-varying polytopic system

$$\dot{x}(t) = A(t)x(t), \quad A(t) \in \hat{\mathscr{A}}_1$$

cannot be proved, not only by using an HPLF of degree 10 but also by using an HPLF of degree 20, although the bounding polytope $\hat{\mathscr{A}}_2$ is only slightly larger than

$\mathscr{A}_2$ (see Figure 3.5). In particular, by using an HPLF of degree 20, the robust stability margin that can be guaranteed with this strategy is equal to 0.6662.

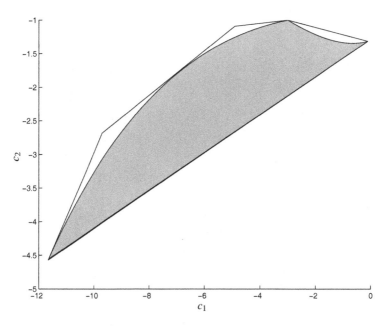

Fig. 3.5 Example HPLF-6: set $\mathscr{A}_2$ (dashed region) and boundary of the polytope $\hat{\mathscr{A}}_2$ (solid line)

3.6   Notes and References

The potential of Lyapunov functions in robustness analysis of uncertain systems affected by parametric uncertainty has been recognized since long time [139, 95]. Robust stability and performance of systems affected by time-varying structured uncertainty has been originally studied via quadratic Lyapunov functions, see for instance [150, 16, 101, 144]. In order to reduce the conservatism, several classes of nonquadratic Lyapunov functions have been intensively investigated, such as piecewise quadratic Lyapunov functions [146, 83, 1, 80], polyhedral Lyapunov functions [17, 6], and HPLFs [19, 148, 32, 82].

The robustness analysis via HPLFs presented in this chapter was proposed in [34] for polytopic systems and in [38] for LFR systems. It is worth mentioning that HPLFs have been exploited in several problems, including the estimation of stability regions for nonlinear systems [62], the evaluation of the minimum dwell time in switching systems [29], and the simultaneous stabilization of linear systems [2]. Piecewise HPLFs have been employed to assess stability of piecewise affine systems [147]. Duality properties of HPLFs have been studied in [69]. Convexity of HPLFs has been investigated in [48].

Chapter 4
Robustness with Time-invariant Uncertainty

Chapter 3 has investigated robust stability and performance of systems with time-varying uncertainties. Another class of fundamental problems in systems engineering concerns robustness analysis of systems affected by time-invariant uncertainties. Such problems arise, for example, when one has to establish whether all systems contained in a given set are stable, or they achieve a guaranteed level of performance. Although the techniques presented in Chapter 3 can be used to address also time-invariant uncertainties, they turn out to be conservative. Actually, the use of parameter-dependent Lyapunov functions can significantly reduce the conservatism of Lyapunov-based robustness analysis techniques.

This chapter describes how robust properties of polytopic systems with time-invariant uncertainties can be investigated by using HPD-QLFs: namely, Lyapunov functions that are quadratic in the state variables and whose dependence on the uncertain parameters is expressed as a matrix form. A key property of this class of Lyapunov functions is that they allow one to formulate robustness conditions in terms of convex optimizations, by adopting the SMR of matrix forms introduced in Chapter 1 and the techniques for characterizing their positivity described in Chapter 2. Another nice feature of HPD-QLFs is that they are non-conservative, in the sense that if a polytopic system is robustly stable, there always exists an HPD-QLF that certifies such a robustness property. Robust stability, as well as evaluation of robust $\mathcal{H}_\infty$ performance and parametric stability margin, are addressed for continuous-time polytopic systems. Extensions to discrete-time systems and to systems affected by rational parametric uncertainty are also presented. *a posteriori* tests for establishing non-conservatism of the results are provided. In particular, it is shown how one can establish whether bounds on robust performance indices are tight, or if a polytopic system is unstable for some values of the uncertain parameters.

G. Chesi et al.: Homogeneous Polynomial Forms, LNCIS 390, pp. 99–132.
springerlink.com   © Springer-Verlag Berlin Heidelberg 2009

4.1 Polytopic Systems with Time-invariant Uncertainty

Let us introduce linear systems affected by time-invariant structured parametric uncertainty.

Definition 4.1 (Time-invariant Polytopic System). Consider the continuous-time system described by

$$\dot{x}(t) = A(p)x(t) \qquad (4.1)$$

where $x(t) \in \mathbb{R}^n$ is the state, $p \in \mathbb{R}^q$ is an uncertain parameter vector supposed to satisfy

$$p \in \Upsilon_q \qquad (4.2)$$

where Υ_q is the simplex in (1.68), and $A(p) \in \mathbb{R}^{n \times n}$ is given by

$$A(p) = \sum_{i=1}^{q} p_i A_i \qquad (4.3)$$

where $A_1, \ldots, A_q \in \mathbb{R}^{n \times n}$ are given matrices. The system (4.1)–(4.3) is called *time-invariant polytopic system*.

In the system (4.1), the vector p represents the time-invariant parametric uncertainty which affects linearly the system dynamics. The vector p can take any value in Υ_q, but it is known to be constant in time. The basic problem addressed in this chapter is to establish whether $A(p)$ is Hurwitz for all admissible values of p, according to the following definition (see Appendix A.2 for definition of Hurwitz matrices).

Definition 4.2 (Robust Stability for Time-invariant Polytopic System). Let us suppose that

$$A \text{ is Hurwitz } \quad \forall A \in \mathscr{A} \qquad (4.4)$$

where

$$\mathscr{A} = \left\{ A(p) \in \mathbb{R}^{n \times n} : \ p \in \Upsilon_q \right\}. \qquad (4.5)$$

Then, $\mathscr{A}$ is said *Hurwitz*, and the system (4.1)–(4.3) is said *robustly stable*.

Therefore, the system (4.1)–(4.3) is robustly stable whenever

$$\mathrm{re}(\lambda) < 0 \quad \forall \lambda \in \mathrm{spc}(A), \ \forall A \in \mathscr{A}.$$

In the sequel the dependence on the time t will be omitted for ease of notation, unless it is required by the context. In order to investigate the robust stability property (4.4), we introduce the following class of Lyapunov functions.

Definition 4.3 (HPD-QLF). Let $v : \mathbb{R}^n \times \Upsilon_q \to \mathbb{R}$ be a function satisfying

$$\begin{cases} v(\cdot, p) \in \Xi_{n,2} & \forall p \in \Upsilon_q \\ v(x, \cdot) \in \Xi_{q,s} & \forall x \in \mathbb{R}^n \\ v(x, p) > 0 & \forall x \in \mathbb{R}_0^n, \ \forall p \in \Upsilon_q \\ \dot{v}(x, p) < 0 & \forall x \in \mathbb{R}_0^n, \ \forall p \in \Upsilon_q \end{cases} \qquad (4.6)$$

where

$$\dot{v}(x,p) = \frac{dv(x,p)}{dt}\bigg|_{\dot{x}=A(p)x}.$$ (4.7)

Then, $v(x,p)$ is said a *HPD-QLF* of degree s for the system (4.1)–(4.3).

Hence, HPD-QLFs are parameter-dependent quadratic Lyapunov functions proving robust stability of the system (4.1)–(4.3) for all admissible values of the uncertainty parameter vector p. The dependence of these functions on p is expressed in terms of forms. In particular, s represents the degree of $v(x,p)$ as a form in p for any fixed x, while the degree of $v(x,p)$ as a form in x is 2 for any fixed p. Clearly, $v(x,p)$ turns out to be a form of degree $s+2$ in both x and p.

HPD-QLFs can be represented in several ways. One possibility is by the vector of coefficients, according to

$$v(x,p) = v'\left(x^{\{2\}} \otimes p^{\{s\}}\right)$$ (4.8)

where $x^{\{2\}} \otimes p^{\{s\}} \in \mathbb{R}^{\sigma(n,2)\sigma(q,s)}$ represents a power vector for the class of HPD-QLFs, and $v \in \mathbb{R}^{\sigma(n,2)\sigma(q,s)}$ is the vector of coefficients of $v(x,p)$ with respect to this power vector. Equivalently, HPD-QLFs can be formulated via the matrix representation

$$v(x,p) = x'V(p)x$$ (4.9)

where $V \in \Xi_{q,s,n}$ is a symmetric matrix form of degree s. The following example illustrates these representations.

Example 4.1. Consider the system (4.1)–(4.3) with $n = 2$, $q = 2$, and

$$A(p) = \begin{pmatrix} 0 & p_1 + p_2 \\ -2p_1 - 2p_2 & -2p_1 - p_2 \end{pmatrix}.$$ (4.10)

Then, the function

$$v(x,p) = (3p_1 + 2p_2)x_1^2 + p_2 x_1 x_2 + (p_1 + p_2)x_2^2$$

is an HPD-QLF for this system, with $s = 1$. This HPD-QLF can be written according to (4.8), with

$$v = \begin{pmatrix} 3 \\ 2 \\ 0 \\ 1 \\ 1 \\ 1 \end{pmatrix}, \quad x^{\{2\}} = \begin{pmatrix} x_1^2 \\ x_1 x_2 \\ x_2^2 \end{pmatrix}, \quad p^{\{1\}} = \begin{pmatrix} p_1 \\ p_2 \end{pmatrix}$$

or according to (4.9), with

$$V(p) = \begin{pmatrix} 3p_1 + 2p_2 & 0.5p_2 \\ \star & p_1 + p_2 \end{pmatrix}.$$

We recall that $V(p)$, being a symmetric matrix form, can be expressed via the SMR as in (1.36). This fact will be exploited in the sequel in order to derive conditions for the existence of HPD-QLFs.

The robust stability property in Definition 4.2 can be investigated via HPD-QLFs. Let us first observe that by expressing $v(x, p)$ as in (4.9), we have that $\dot{v}(x, p)$ can be written as

$$\dot{v}(x, p) = x' \, he \, (V(p)A(p)) \, x. \tag{4.11}$$

The following result states that HPD-QLFs are non-conservative for establishing whether the system (4.1)–(4.3) is robustly stable.

Theorem 4.1. *Let us define*

$$\mu = \frac{1}{2}n(n+1) - 1. \tag{4.12}$$

The system (4.1)–(4.3) *is robustly stable if and only if there exists a matrix form* $V \in \Xi_{q,s,n}$, *with* $s \le \mu$, *such that*

$$\begin{cases} 0 < V(p) \\ 0 > he \, (V(p)A(p)) \end{cases} \forall p \in \Upsilon_q. \tag{4.13}$$

Proof. (Sufficiency) Let us suppose that there exists $V \in \Xi_{q,s,n}$ such that (4.13) holds, and let us define $v(x, p)$ as in (4.9). By taking into account (4.11), it follows that $v(x, p)$ satisfies the condition (4.6), and hence $\mathscr{A}$ is Hurwitz.

(Necessity) Let us suppose that the system (4.1)–(4.3) is robustly stable, and let $\tilde{E} \in \Xi_{q,\tilde{s},n}$ be any symmetric matrix form of degree $\tilde{s}$ such that

$$\tilde{E}(p) > 0 \ \forall p \in \Upsilon_q$$

and consider the Lyapunov equation

$$he \, (\tilde{V}(p)A(p)) = -\tilde{E}(p). \tag{4.14}$$

Being $\mathscr{A}$ Hurwitz, the solution of (4.14) is a symmetric matrix function $\tilde{V}(p)$ which is positive definite for all $p \in \Upsilon_q$ and which depends rationally on p. Indeed, by stacking the $n(n+1)/2$ entries of $\tilde{V}(p)$ and $\tilde{E}(p)$ into the vectors $\tilde{v}(p)$ and $\tilde{e}(p)$, respectively, one can rewrite (4.14) as

$$M(p)\tilde{v}(p) = \tilde{e}(p) \tag{4.15}$$

where the entries of $M(p)$ are linear in p. Therefore, the entries of the solution $\tilde{v}(p)$ of (4.15) can be expressed as

$$\tilde{v}_j(p) = \frac{\tilde{n}_j(p)}{d(p)}, \quad j = 1, \ldots, \frac{1}{2}n(n+1)$$

where $\tilde{n}_j(p)$ and $d(p)$ are forms whose degree is at most $n(n+1)/2 - 1 + \tilde{s}$ and $n(n+1)/2$, respectively. Moreover, being $\tilde{V}(p) > 0$ for all $p \in \Upsilon_q$, one has that $d(p) > 0$ for all $p \in \Upsilon_q$. By defining

$$V(p) = d(p)\tilde{V}(p) \tag{4.16}$$

one has that $V(p)$ satisfies (4.13), in particular

$$\text{he}\,(V(p)A(p)) = -d(p)\tilde{E}(p).$$

Finally, by (4.16) the degree s of $V(p)$ is the same as that of the forms $\tilde{n}_j(p)$, and hence

$$s \leq \frac{1}{2}n(n+1) - 1 + \tilde{s}.$$

Since $\tilde{s}$ is arbitrary, one can select $\tilde{s} = 0$, and hence one gets $s \leq \mu$ with μ given by (4.12). □

4.2 Robust Stability

In this section, conditions for robust stability of the system (4.1)–(4.3) are provided in terms of LMI problems.

4.2.1 Parametrization of HPD-QLFs

The aim is to find an HPD-QLF as in Definition 4.3, described by the matrix representation (4.9), with $V(p)$ satisfying conditions (4.13) in Theorem 4.1. The first condition to be enforced is the positive definiteness of the HPD-QLF matrix $V(p)$ within the set Υ_q, i.e. the first inequality in (4.13). In this respect, a parametrization of positive matrix forms $V(p)$, for p belonging to Υ_q, is provided next. The second inequality in (4.13) will be addressed in Section 4.2.2.

From Theorem 1.17 one has that $V(p)$ is positive definite for all $p \in \Upsilon_q$ if and only if $V(\text{sq}(p))$ is positive definite for all $p \in \mathbb{R}_0^q$. Let us observe that $V(\text{sq}(p))$ can be written by using the notation introduced in (1.35) as

$$V(\text{sq}(p)) = \Phi\left(S, p^{\{s\}}, n\right) \tag{4.17}$$

for some suitable matrix $S \in \mathscr{S}_{q,s,n}$ where

$$\mathscr{S}_{q,s,n} = \Big\{ S \in \mathbb{S}^{n\sigma(q,s)} : \; \Phi\left(S, p^{\{s\}}, n\right) \text{ does not}$$
$$\text{contain monomials } p_1^{i_1} p_2^{i_2} \dots p_q^{i_q} \text{ with any odd } i_j \Big\}. \tag{4.18}$$

In fact, it is straightforward to verify that (4.17) holds for some symmetric matrix S if and only if this matrix S belongs to the set $\mathscr{S}_{q,s,n}$. Hence, due to (4.17), one has that

$$S > 0 \;\Rightarrow\; V(\mathrm{sq}(p)) > 0 \;\; \forall p \in \mathbb{R}_0^q.$$

Moreover, due to Theorem 1.17, one has that

$$S > 0 \;\Rightarrow\; V(p) > 0 \;\; \forall p \in \Upsilon_q.$$

In order to increase the degrees of freedom in the selection of $V(p)$, one can exploit the fact that the matrix S in (4.17) is not unique. Indeed, the following result provides a characterization of the set $\mathscr{S}_{q,s,n}$.

Theorem 4.2. *The set $\mathscr{S}_{q,s,n}$ is a linear space of dimension*

$$\tau(q,s,n) = \frac{1}{2} n \left(\sigma(q,s)(n\sigma(q,s)+1) - (n+1)(\sigma(q,2s) - \sigma(q,s)) \right). \tag{4.19}$$

Proof. It is immediate to check that if $S_1, S_2 \in \mathscr{S}_{q,s,n}$, then $a_1 S_1 + a_2 S_2 \in \mathscr{S}_{q,s,n}$ for all $a_1, a_2 \in \mathbb{R}$. Then, let us observe that $n\sigma(q,s)(n\sigma(q,s)+1)/2$ is the number of distinct entries of a symmetric matrix of dimension $n\sigma(q,s) \times n\sigma(q,s)$, whereas $n(n+1)(\sigma(q,2s) - \sigma(q,s))/2$ is the total number of monomials in $\Phi\left(S, p^{\{s\}}, n\right)$ containing at least one odd power. The constraints obtained by annihilating these monomials are linear and independent, similarly to the proof of Theorem 1.2. Therefore, the dimension of $\mathscr{S}_{q,s,n}$ is given by $\tau(q,s,n)$ in (4.19). $\qquad\square$

The following result summarizes the strategy for generating matrix forms $V(p)$ satisfying the first inequality in (4.13), and stems directly from the definition of $\mathscr{S}_{q,s,n}$ and from Theorem 1.17.

Theorem 4.3. *Let $S(\beta)$ be a linear parametrization of $\mathscr{S}_{q,s,n}$ in (4.18), with $\beta \in \mathbb{R}^{\tau(q,s,n)}$. Let $V(p;\beta)$ be the parametrized matrix form defined by*

$$V(\mathrm{sq}(p);\beta) = \Phi\left(S(\beta), p^{\{s\}}, n\right). \tag{4.20}$$

Then,

$$V \in \Xi_{q,s,n} \iff \exists \beta : V(p) = V(p;\beta).$$

Moreover,

$$\exists \beta : S(\beta) > 0 \;\Rightarrow\; V(p;\beta) > 0 \;\; \forall p \in \Upsilon_q.$$

4.2.2 Conditions for Robust Stability

In the following, a condition for the solution of the robust stability problem is provided by enforcing the second condition in (4.13). To this purpose, let us define

$$Q(p;\beta) = \text{he}(V(p;\beta)A(p)). \qquad (4.21)$$

We have that $Q(\cdot;\beta) \in \Xi_{q,s+1,n}$ for all β, i.e. $Q(p;\beta)$ is a matrix form of degree $s+1$ in $p \in \mathbb{R}^q$, linearly parametrized by β. Let us express $Q(\text{sq}(p);\beta)$ as

$$Q(\text{sq}(p);\beta) = \Phi\left(R(\beta), p^{\{s+1\}}, n\right) \qquad (4.22)$$

where $R(\beta) \in \mathbb{S}^{n\sigma(q,s+1)}$ is an SMR matrix of $Q(\text{sq}(p);\beta)$. Let us observe that $R(\beta)$ depends linearly on β, because $Q(\text{sq}(p);\beta)$ depends linearly on $V(\text{sq}(p);\beta)$, which in turn depends linearly on β. The following result yields a condition for robust stability of the system (4.1)–(4.3) via HPD-QLFs.

Theorem 4.4. *The system* (4.1)–(4.3) *is robustly stable if there exist an integer* $s \geq 0$, $\alpha \in \mathbb{R}^{\omega(q,s+1,n)}$ *and* $\beta \in \mathbb{R}^{\tau(q,s,n)}$ *such that*

$$\begin{cases} 0 < S(\beta) \\ 0 > R(\beta) + L(\alpha) \end{cases} \qquad (4.23)$$

where $S(\beta)$ *is a linear parametrization of* $\mathscr{S}_{q,s,n}$ *in* (4.18), $L(\alpha)$ *is a linear parametrization of* $\mathscr{L}_{q,s+1,n}$ *in* (1.39), *and* $R(\beta)$ *is defined according to* (4.21)–(4.22).

Proof. Let α and β be such that (4.23) holds. Let us select $V \in \Xi_{q,s,n}$ as

$$V(p) = V(p;\beta)$$

with $V(p;\beta)$ defined according to (4.20). From the first inequality in (4.23) and Theorem 4.3 one has that $V(p) > 0$ for all $p \in \Upsilon_q$, and hence the first condition in (4.13) holds. Then, by (4.21)–(4.22) one has that

$$\Phi\left(R(\beta) + L(\alpha), p^{\{s+1\}}, n\right) = \Phi\left(R(\beta), p^{\{s+1\}}, n\right)$$
$$= \text{he}(V(\text{sq}(p))A(\text{sq}(p))).$$

By the second inequality in (4.23), this implies that

$$\text{he}(V(\text{sq}(p))A(\text{sq}(p))) < 0 \quad \forall p \in \mathbb{R}_0^q.$$

As a consequence, one has from Theorem 1.17 that $\text{he}(V(p)A(p)) < 0$ for all $p \in \Upsilon_q$, and hence also the second condition in (4.13) holds. Therefore, the system is robustly stable. $\qquad \Box$

Theorem 4.4 provides a condition for robust stability of the system (4.1)–(4.3) which amounts to solving an LMI feasibility test. In particular, the number of scalar variables in (4.23) is given by

$$\tau(q,s,n) + \omega(q,s+1,n)$$

where $\tau(\cdot,\cdot,\cdot)$ and $\omega(\cdot,\cdot,\cdot)$ are defined in (4.19) and (1.40), respectively. Table 4.1 shows this number for some values of n,s,q.

Table 4.1 Number of scalar variables in the LMI feasibility test (4.23) for some values of n,s,q: (a) $q=2$; (b) $q=3$

	(a)					(b)			
	$s=0$	$s=1$	$s=2$	$s=3$		$s=0$	$s=1$	$s=2$	$s=3$
$n=2$	4	13	30	55	$n=2$	6	45	177	486
$n=3$	9	30	69	126	$n=3$	15	108	414	1122
$n=4$	16	54	124	226	$n=4$	28	198	750	2020

A question that naturally arises is whether there exists a relationship between families of HPD-QLFs of different degrees. The following result clarifies that the conservatism of the condition in Theorem 4.4 does not increase with s.

Theorem 4.5. *If condition (4.23) of Theorem 4.4 holds for some integer $s \geq 0$, then it holds also for $s+1$.*

Proof. Let us first rewrite (4.23) with s replaced by $s+1$ as

$$\begin{cases} 0 < \tilde{S}(\tilde{\beta}) \\ 0 > \tilde{R}(\tilde{\beta}) + \tilde{L}(\tilde{\alpha}) \end{cases} \qquad (4.24)$$

where $\tilde{S}(\tilde{\beta})$ is a linear parametrization of $\mathscr{S}_{q,s+1,n}$ with $\tilde{\beta} \in \mathbb{R}^{s(q,s+1,n)}$; $\tilde{L}(\tilde{\alpha})$ is a linear parametrization of $\mathscr{L}_{q,s+2,n}$ with $\tilde{\alpha} \in \mathbb{R}^{\omega(q,s+2,n)}$; $\tilde{R}(\tilde{\beta})$ is a linear matrix function such that

$$\tilde{Q}(\mathrm{sq}(p);\tilde{\beta}) = \Phi\left(\tilde{R}(\tilde{\beta}), p^{\{s+2\}}, n\right) \qquad (4.25)$$

with

$$\tilde{Q}(p;\tilde{\beta}) = \mathrm{he}\left(\tilde{V}(p;\tilde{\beta})A(p)\right) \qquad (4.26)$$

$$\tilde{V}(\mathrm{sq}(p);\tilde{\beta}) = \Phi\left(\tilde{S}(\tilde{\beta}), p^{\{s+1\}}, n\right). \qquad (4.27)$$

Let α and β be such that (4.23) holds. From the proof of Theorem 4.4 we have that there exists $V \in \Xi_{q,s,n}$ such that $V(p) > 0$ and $\mathrm{he}\,(V(p)A(p)) < 0$ for all $p \in \Upsilon_q$. Let us define now

$$\tilde{V}(p) = V(p) \sum_{i=1}^{q} p_i.$$

It follows that

$$\tilde{V}(p) > 0 \quad \forall p \in \Upsilon_q.$$

Moreover,

$$\text{he}\left(\tilde{V}(p)A(p)\right) = \text{he}\left(V(p)A(p)\right) \sum_{i=1}^{q} p_i$$

and hence

$$\text{he}\left(\tilde{V}(p)A(p)\right) < 0 \quad \forall p \in \Upsilon_q.$$

This implies that (4.13) is satisfied also when s is replaced by $s+1$, i.e. $\tilde{v}(x,p) = x'\tilde{V}(p)x$ is an HPD-QLF of degree $s+1$.

Now, let us show that $\tilde{V}(\text{sq}(p))$ admits a positive definite SMR matrix, i.e. there exists $\tilde{S}(\tilde{\beta}) > 0$ such that (4.27) holds. Let T_{s+1} be the matrix satisfying

$$p \otimes p^{\{s\}} = T_{s+1} p^{\{s+1\}} \quad \forall p \in \mathbb{R}^q. \tag{4.28}$$

Then,

$$\begin{aligned}
\tilde{V}(\text{sq}(p)) &= \left(\sum_{i=1}^{q} p_i^2\right) \Phi\left(S(\beta), p^{\{s\}}, n\right) \\
&= p'p\Phi\left(S(\beta), p^{\{s\}}, n\right) \\
&= \Phi\left(I_q \otimes S(\beta), p \otimes p^{\{s\}}, n\right) \\
&= \Phi\left(I_q \otimes S(\beta), T_{s+1} p^{\{s+1\}}, n\right) \\
&= \Phi\left(\check{S}, p^{\{s+1\}}, n\right)
\end{aligned} \tag{4.29}$$

where

$$\begin{aligned}
\check{S} &= (T_{s+1} \otimes I_n)' \left(I_q \otimes S(\beta)\right) (T_{s+1} \otimes I_n) \\
&= \Phi\left(I_q \otimes S(\beta), T_{s+1}, n\right)
\end{aligned}$$

and the latter equality exploits (with a slight abuse of notation) the parametrization of symmetric matrix forms introduced in (1.35). Since $S(\beta) > 0$ and T_{s+1} is a matrix with full column rank, it follows that $\check{S} > 0$. Moreover, from (4.29) one has that $\check{S} \in \mathscr{S}_{q,s+1,n}$, and hence there exists $\tilde{\beta}$ such that $\tilde{S}(\tilde{\beta}) = \check{S}$ in (4.27).

Let us now turn the attention to the second inequality in (4.24). We want to show that $\text{he}\left(\tilde{V}(\text{sq}(p))A(\text{sq}(p))\right)$ admits a negative definite SMR matrix, i.e. (4.25) holds for some $\tilde{R}(\tilde{\beta})$ satisfying (4.24). By following the same development as in (4.29), one gets

$$\text{he}\left(\tilde{V}(\text{sq}(p))A(\text{sq}(p))\right) = \Phi\left(\check{R}, p^{\{s+2\}}, n\right)$$

where

$$\check{R} = \Phi\left(I_q \otimes (R(\beta) + L(\alpha)), T_{s+2}, n\right).$$

The result follows by observing that $\tilde{R}(\tilde{\beta})$ in (4.25) and $\check{R}$ are SMR matrices of the same matrix form. Therefore, there exists $\tilde{\alpha}$ such that $\tilde{R}(\tilde{\beta}) + \tilde{L}(\tilde{\alpha}) = \check{R}$, and the second inequality in (4.24) holds since $R(\beta) + L(\alpha) < 0$ and T_{s+2} is a matrix with full column rank. $\square$

The following result states that the condition for robust stability provided by Theorem 4.4 is not only sufficient but also necessary for some sufficiently large integer s.

Theorem 4.6. *Let us suppose that the system* (4.1)–(4.3) *is robustly stable. Then, there exists an integer* $v \geq 0$ *such that condition* (4.23) *of Theorem 4.4 holds for* $s = v$.

Proof. Let us suppose that $\mathscr{A}$ is Hurwitz. Then, from Theorem 4.1 there exists $V \in \Xi_{q,s,n}$ such that (4.13) is satisfied. By using Theorem 1.17 this implies that $V(p)$ satisfies also

$$\begin{cases} 0 < V(\mathrm{sq}(p)) \\ 0 > \mathrm{he}\,(V(\mathrm{sq}(p))A(\mathrm{sq}(p))) \end{cases} \quad \forall p \in \mathbb{R}_0^q. \tag{4.30}$$

Pre- and post-multiplying the inequalities in (4.30) by x' and x respectively, we obtain

$$\begin{cases} 0 < v(x, \mathrm{sq}(p)) \\ 0 > \dot{v}(x, \mathrm{sq}(p)) \end{cases} \quad \forall x \in \mathbb{R}_0^n, \; \forall p \in \mathbb{R}_0^q. \tag{4.31}$$

Moreover, since $v(x, \mathrm{sq}(p))$ is positive definite and $\dot{v}(x, \mathrm{sq}(p))$ is negative definite, by a trivial extension of Theorem 1.8 it follows that there exist $\varepsilon_1 > 0$ and $\varepsilon_2 > 0$ such that

$$\begin{cases} 0 < v(x, \mathrm{sq}(p)) - \varepsilon_1 \|x\|^2 \|p\|^{2s} \\ 0 > \dot{v}(x, \mathrm{sq}(p)) + \varepsilon_2 \|x\|^2 \|p\|^{2(s+1)} \end{cases} \quad \forall x \in \mathbb{R}_0^n, \; \forall p \in \mathbb{R}_0^q. \tag{4.32}$$

From Theorem 2.1 we have that any positive semidefinite form can be written as the ratio of two SOS forms. Hence, condition (4.32) holds if and only if there exist SOS forms $z_1(x, \mathrm{sq}(p)), \ldots, z_4(x, \mathrm{sq}(p))$ such that

$$\begin{aligned} v(x, \mathrm{sq}(p)) - \varepsilon_1 \|x\|^2 \|p\|^{2s} &= \frac{z_1(x, \mathrm{sq}(p))}{z_2(x, \mathrm{sq}(p))} \\ \dot{v}(x, \mathrm{sq}(p)) + \varepsilon_2 \|x\|^2 \|p\|^{2(s+1)} &= -\frac{z_3(x, \mathrm{sq}(p))}{z_4(x, \mathrm{sq}(p))}. \end{aligned} \tag{4.33}$$

Now, $v(x, \mathrm{sq}(p)) - \varepsilon_1 \|x\|^2 \|p\|^{2s}$ is quadratic in x, and any quadratic form is positive if and only if it is SOS, by Theorem 2.3. Hence,

$$v(x, \mathrm{sq}(p)) - \varepsilon_1 \|x\|^2 \|p\|^{2s} = \sum_{i=1}^{n_0} \left(a_i(p)'x \right)^2 \tag{4.34}$$

for some functions $a_i(p) \in \mathbb{R}^n$. Therefore, from (4.33) and (4.34) we can conclude that $z_1(x, \mathrm{sq}(p))$ and $z_2(x, \mathrm{sq}(p))$ can be selected quadratic and constant in x, respectively. Moreover, $v(x, p) - \varepsilon_1 \|x\|^2 \|p\|^s$ is positive definite for all $p \in \Upsilon$, which implies from Theorem 2.2 that $z_2(x, p)$ can be selected as a form with positive coefficients. The same reasoning applies also to $\dot{v}(x, \mathrm{sq}(p)) + \varepsilon_2 \|x\|^2 \|p\|^{2(s+1)}$, hence implying that $z_1(x, \mathrm{sq}(p)), \ldots, z_4(x, \mathrm{sq}(p))$ can be expressed as

$$z_j(x, \mathrm{sq}(p)) = \sum_{i=1}^{n_j} \left(b_{j,i}(p)'x \right)^2, \ \ j = 1, 3$$

$$z_j(x, \mathrm{sq}(p)) = b_j(\mathrm{sq}(p)), \ \ j = 2, 4$$

$$b_j(\mathrm{sq}(p)) = p^{\{\delta_j\}'} D_j p^{\{\delta_j\}}, \ \ j = 2, 4$$

where $\delta_1, \ldots, \delta_4$ are nonnegative integers, $b_{1,i}(p)$ and $b_{3,i}(p)$ are forms of degree δ_1 and δ_3 respectively, and D_2 and D_4 are diagonal positive definite matrices.

Let us define the function

$$\tilde{v}(x, \mathrm{sq}(p)) = b_2(\mathrm{sq}(p)) b_4(\mathrm{sq}(p)) v(x, \mathrm{sq}(p)) \tag{4.35}$$

of degree $2v$ in p, where

$$v = s + \delta_2 + \delta_4.$$

This function is an HPD-QLF because

$$\begin{cases} 0 < \tilde{v}(x, \mathrm{sq}(p)) \\ 0 > \dot{\tilde{v}}(x, \mathrm{sq}(p)) \end{cases} \quad \forall x \in \mathbb{R}_0^n, \ \forall p \in \mathbb{R}_0^q$$

where the first inequality is obvious and the second one holds due to

$$\dot{\tilde{v}}(x, \mathrm{sq}(p)) = b_2(\mathrm{sq}(p)) b_4(\mathrm{sq}(p)) \dot{v}(x, \mathrm{sq}(p))$$

and (4.31). Moreover, $\tilde{v}(x, \mathrm{sq}(p))$ and $\dot{\tilde{v}}(x, \mathrm{sq}(p))$ can be expressed as

$$\tilde{v}(x, \mathrm{sq}(p)) = x' \tilde{V}(\mathrm{sq}(p)) x$$
$$\dot{\tilde{v}}(x, \mathrm{sq}(p)) = x' \tilde{Q}(\mathrm{sq}(p)) x$$

where

$$\tilde{V}(\mathrm{sq}(p)) = \varepsilon_1 c(\mathrm{sq}(p)) I_n + b_4(\mathrm{sq}(p)) \sum_{i=1}^{n_1} b_{1,i}(p) b_{1,i}(p)' \tag{4.36}$$

$$\tilde{Q}(\mathrm{sq}(p)) = \varepsilon_2 \|p\|^2 c(\mathrm{sq}(p)) I_n + b_2(\mathrm{sq}(p)) \sum_{i=1}^{n_3} b_{3,i}(p) b_{3,i}(p)' \tag{4.37}$$

$$c(\mathrm{sq}(p)) = \|p\|^{2s} b_2(\mathrm{sq}(p)) b_4(\mathrm{sq}(p)). \tag{4.38}$$

From (4.36)–(4.38) we have that

$$\tilde{V}(\text{sq}(p)) = \varepsilon_1 c(\text{sq}(p)) I_n + \sum_{i=1}^{n_5} b_{5,i}(p) b_{5,i}(p)'$$

$$\tilde{Q}(\text{sq}(p)) = \varepsilon_2 \|p\|^2 c(\text{sq}(p)) I_n + \sum_{i=1}^{n_6} b_{6,i}(p) b_{6,i}(p)'$$

where each entry of the vectors $b_{5,i}(p)$ and $b_{6,i}(p)$ is a form. Then, by expressing $b_{5,i}(p)$ and $b_{6,i}(p)$ as

$$b_{5,i}(p) = \hat{b}_{5,i}' \left(p^{\{v\}} \otimes I_n \right)$$
$$b_{6,i}(p) = \hat{b}_{6,i}' \left(p^{\{v+1\}} \otimes I_n \right)$$

where $\hat{b}_{5,i}$ and $\hat{b}_{6,i}$ are suitable vectors, we get

$$\begin{aligned}
\tilde{V}(\text{sq}(p)) &= \Phi\left(\tilde{S}, p^{\{v\}}, n \right) \\
\tilde{Q}(\text{sq}(p)) &= \Phi\left(\tilde{R}, p^{\{v+1\}}, n \right)
\end{aligned} \tag{4.39}$$

where

$$\tilde{S} = \varepsilon_1 \left(F_1 \otimes I_n \right) + \sum_{i=1}^{n_5} \hat{b}_{5,i} \hat{b}_{5,i}'$$

$$\tilde{R} = \varepsilon_2 \left(F_2 \otimes I_n \right) + \sum_{i=1}^{n_6} \hat{b}_{6,i} \hat{b}_{6,i}' \tag{4.40}$$

$$F_1 = K_v' \left(E_s \otimes D_2 \otimes D_4 \right) K_v$$
$$F_2 = K_{v+1}' \left(E_{s+1} \otimes D_2 \otimes D_4 \right) K_{v+1}$$

where K_v and K_{v+1} are defined by (3.9) for x replaced by p, and E_δ is the diagonal positive definite matrix satisfying

$$p^{\{\delta\}'} E_\delta p^{\{\delta\}} = \|p\|^{2\delta}.$$

Hence, $F_1 > 0$, $F_2 > 0$, and therefore from (4.40) it follows that $\tilde{S} > 0$ and $\tilde{R} > 0$. Finally, $\tilde{S} \in \mathscr{S}_{q,v,n}$, and hence there exists β such that $S(\beta) = \tilde{S}$ where $S(\beta)$ is a linear parametrization of $\mathscr{S}_{q,v,n}$. Similarly, for any SMR matrix $R(\beta)$ of $\tilde{Q}(\text{sq}(p))$, by (4.39) one has

$$\Phi\left(R(\beta) - \tilde{R}, p^{\{v+1\}}, n \right) = 0_{n \times n}$$

and hence there exists α such that $R(\beta) + L(\alpha) = \tilde{R}$, where $L(\alpha)$ is a linear parametrization of $\mathscr{L}_{q,v+1,n}$. Therefore, Theorem 4.4 is satisfied with $s = v$. $\square$

4.2.3 Detecting Instability

In the previous sections it has been shown how a sufficient condition for robust stability of the system (4.1)–(4.3) can be formulated as an LMI feasibility test, and that this condition is also necessary for some degree of the HPD-QLF.

Now, let us suppose that the system is unstable for some uncertainty vector p. Then, the LMI condition will be never satisfied, and since the degree of the HPD-QLF required to achieve necessity is unknown and can be arbitrarily high, then one can not conclude that the system is not robustly stable. The following result proposes a strategy to address this problem.

Theorem 4.7. *Let us suppose that the system* (4.1)–(4.3) *is not robustly stable, and let us define the EVP*

$$z^* = \sup_{z \in \mathbb{R}, \ \beta \in \mathbb{R}^{\tau(q,s,n)}, \ \alpha \in \mathbb{R}^{\omega(q,s+2,n)}} z \tag{4.41}$$

$$s.t. \quad \begin{cases} 0 < S(\beta) \\ 0 > R(\beta) + L(\alpha) + zI \end{cases}$$

where $S(\beta)$ *is a linear parametrization of* $\mathscr{S}_{q,s,n}$ *in* (4.18), $L(\alpha)$ *is a linear parametrization of* $\mathscr{L}_{q,s+1,n}$ *in* (1.39), *and* $R(\beta)$ *is defined according to* (4.21)–(4.22). *Let*

$$M = R(\beta^*) + L(\alpha^*) \tag{4.42}$$

where β^*, α^* *are optimal values of* β, α *in* (4.41). *Then, there exist* $\hat{p} \in \Upsilon_q$ *and* $\hat{x} \in \mathbb{R}_0^n$ *such that* $A(\hat{p})$ *is not Hurwitz and*

$$\left(\mathrm{sqr}(\hat{p})^{\{s+1\}} \otimes \hat{x}\right)' M \left(\mathrm{sqr}(\hat{p})^{\{s+1\}} \otimes \hat{x}\right) \geq 0. \tag{4.43}$$

Proof. Let us define

$$V(p) = \Phi\left(S(\beta^*), \mathrm{sqr}(p)^{\{s\}}, n\right).$$

It follows that

$$\mathrm{he}\,(V(p)A(p)) = \Phi\left(M, \mathrm{sqr}(p)^{\{s+1\}}, n\right). \tag{4.44}$$

Let $\hat{p}$ be any vector in Υ_q such that $A(\hat{p})$ is not Hurwitz. Since $S(\beta^*) > 0$ it follows that $V(\hat{p}) > 0$, which in turn implies $\mathrm{he}\,(V(\hat{p})A(\hat{p})) \not< 0$. This means that there exists $x \in \mathbb{R}_0^n$ such that

$$\hat{x}'\,\mathrm{he}\,(V(\hat{p})A(\hat{p}))\,\hat{x} \geq 0.$$

Taking into account (4.44) we have

$$0 \leq \hat{x}'\Phi\left(M, \mathrm{sqr}(\hat{p})^{\{s+1\}}, n\right)\hat{x}$$

$$= \left(\mathrm{sqr}(\hat{p})^{\{s+1\}} \otimes \hat{x}\right)' M \left(\mathrm{sqr}(\hat{p})^{\{s+1\}} \otimes \hat{x}\right)$$

and hence the theorem holds. $\qquad\square$

Theorem 4.7 characterizes the situation in which the system (4.1)–(4.3) is not robustly stable, in terms of the matrix M in (4.43), which is obtained by solving (4.41). Let us observe that a necessary condition for (4.43) to hold is that $\lambda_{max}(M) \geq 0$, or equivalently z^* in (4.41) satisfies

$$z^* \leq 0.$$

Then, one has to look for vectors $\hat{p} \in Y_q$ and $\hat{x} \in \mathbb{R}_0^n$ such that (4.43) holds. This is not an easy problem, as it amounts to detecting values of the variables that fulfill a polynomial inequality. Nevertheless, a simpler instability test can be formulated by restricting the set of admissible vectors $\mathrm{sqr}(\hat{p})^{\{s+1\}} \otimes \hat{x}$ to a given domain, e.g. a linear space. In this respect, a possible choice is the eigenspace of the largest eigenvalue of M. Hence, the problem becomes to find the vectors $\hat{p} \in Y_q$ and $\hat{x} \in \mathbb{R}_0^n$ such that

$$\left(\mathrm{sqr}(\hat{p})^{\{s+1\}} \otimes \hat{x} \right) \in \ker(M + z^*I). \tag{4.45}$$

Let us denote the set of such vectors $\hat{p}$ as

$$\mathscr{P}_M = \left\{ \hat{p} \in Y_q : \text{(4.45) holds for some } \hat{x} \in \mathbb{R}_0^n \right\}. \tag{4.46}$$

Then, one may look for vectors $\hat{p}$ such that $A(\hat{p})$ is not Hurwitz, within the set of candidates $\mathscr{P}_M$. Strategies for finding a solution of problem (4.45) will be proposed in Section 4.3.1 (see Theorem 4.11 and the subsequent discussion).

4.3 Robust Performance

In this section, we use HPD-QLFs to tackle some robust performance problems involving systems affected by time-invariant parametric uncertainty.

4.3.1 Robust $\mathscr{H}_\infty$ Performance

In the following, we investigate the use of HPD-QLFs for evaluating robust $\mathscr{H}_\infty$ performance. Specifically, let us consider the time-invariant polytopic system

$$\begin{cases} \dot{x}(t) = A(p)x(t) + B(p)w(t) \\ y(t) = C(p)x(t) + D(p)w(t) \\ p \in Y_q \end{cases} \tag{4.47}$$

where $x \in \mathbb{R}^n$ is the state, $w \in \mathbb{R}^r$ is the input, $y \in \mathbb{R}^g$ is the output, and $p \in \mathbb{R}^q$ is the parametric uncertainty vector. The matrices $A(p) \in \mathbb{R}^{n \times n}$, $B(p) \in \mathbb{R}^{n \times r}$, $C(p) \in \mathbb{R}^{g \times n}$ and $D(p) \in \mathbb{R}^{g \times r}$ are linear functions of p according to

$$A(p) = \sum_{i=1}^{q} p_i A_i, \quad B(p) = \sum_{i=1}^{q} p_i B_i$$
$$C(p) = \sum_{i=1}^{q} p_i C_i, \quad D(p) = \sum_{i=1}^{q} p_i D_i \tag{4.48}$$

where $A_i \in \mathbb{R}^{n \times n}, B_i \in \mathbb{R}^{n \times r}, C_i \in \mathbb{R}^{g \times n}, D_i \in \mathbb{R}^{g \times r}, i = 1, \ldots, q$, are given matrices. For any $p \in \Upsilon_q$, the transfer function from w to y is given by

$$H(\delta, p) = C(p) (\delta I_n - A(p))^{-1} B(p) + D(p).$$

For a fixed p, the $\mathcal{H}_\infty$ norm of $H(\delta, p)$ is given by

$$\|H(\cdot, p)\|_\infty = \sup_{\omega \in \mathbb{R}} \|H(j\omega, p)\|_\infty.$$

This norm can be computed through the bounded real lemma in the following way (see for instance [14]):

$$\|H(\cdot, p)\|_\infty = \inf_{\gamma \in \mathbb{R}, V \in \mathbb{S}^n} \gamma$$
$$\text{s.t.} \begin{cases} 0 < V \\ 0 > E(V, p) + \dfrac{1}{\gamma^2} F(p) \end{cases} \tag{4.49}$$

where

$$E(V, p) = \begin{pmatrix} \text{he}(VA(p)) & VB(p) \\ \star & -I_r \end{pmatrix} \tag{4.50}$$

$$F(p) = \begin{pmatrix} C(p)' \\ D(p)' \end{pmatrix} \begin{pmatrix} C(p) & D(p) \end{pmatrix}. \tag{4.51}$$

The problem we address is the computation of the worst-case $\mathcal{H}_\infty$ norm of system (4.47)–(4.48) defined hereafter.

Definition 4.4 (Robust $\mathcal{H}_\infty$ Performance). Let us define

$$\gamma^{H_\infty} = \sup_{p \in \Upsilon_q} \|H(\cdot, p)\|_\infty. \tag{4.52}$$

Then, γ^{H_∞} is the *robust $\mathcal{H}_\infty$ performance* of the system (4.47)–(4.48).

An upper bound to γ^{H_∞} can be computed by employing a common quadratic Lyapunov function. This can be done by solving an EVP like (4.49), where the second LMI constraint is enforced in all the vertices of the simplex Υ_q (see [14, pp. 91-92]). Unfortunately, such an upper bound is generally very conservative. It is well-known that one can reduce the conservatism by employing parameter-dependent Lyapunov functions. However, direct use of (4.49) is not possible anymore, because the problem turns out to be nonconvex, and therefore suitable convex relaxations must be devised.

In the following, we tackle the computation of the worst-case $\mathcal{H}_\infty$ norm by employing HPD-QLFs, such as $v(x, p) = x'V(p)x$. Being such Lyapunov functions quadratic in the state, (4.49) can be immediately extended to HPD-QLFs by replacing V by the matrix form $V(p)$. Specifically, we define the $\mathcal{H}_\infty$ cost guaranteed by HPD-QLFs of a given degree s as follows.

Definition 4.5 (s-HPD-QLF Robust $\mathcal{H}_\infty$ Performance). Let us define

$$
\gamma_s^{H_\infty} = \inf_{\gamma \in \mathbb{R}, \, V \in \Xi_{q,s,n}} \gamma
$$

$$
\text{s.t.} \quad \begin{cases} 0 < V(p) \\ 0 > E(V(p),p) + \dfrac{1}{\gamma^2}F(p) \end{cases} \quad \forall p \in \Upsilon_q. \tag{4.53}
$$

Then, $\gamma_s^{H_\infty}$ is said s-HPD-QLF robust $\mathcal{H}_\infty$ performance of the system (4.47)–(4.48).

Let us observe that $\gamma_s^{H_\infty}$ is an upper bound of γ^{H_∞} for any possible s, i.e.

$$
\gamma_s^{H_\infty} \geq \gamma^{H_\infty} \quad \forall s \geq 0.
$$

However, (4.53) is not a convex problem and we need to introduce a convex relaxation. Hereafter, it is shown how upper bounds to $\gamma_s^{H_\infty}$ can be computed through convex optimizations, by employing the same machinery used in Section 4.2 to study robust stability problems. To this purpose, let $V(p;\beta)$ be defined by (4.20), and let us introduce

$$
Q(p;\beta,z) = E(V(p;\beta),p) + zF(p)\left(\sum_{i=1}^{q} p_i\right)^{s-1} + G(p) \tag{4.54}
$$

where $\beta \in \mathbb{R}^{\tau(q,s,n)}$, $z \in \mathbb{R}$, $V(p;\beta)$ is defined by (4.20), and

$$
G(p) = \begin{pmatrix} 0_{n\times n} & 0_{n\times r} \\ \star & \left(1 - \left(\sum_{i=1}^{q} p_i\right)^{s+1}\right)I_r \end{pmatrix}. \tag{4.55}
$$

We have that $Q(\cdot;\beta,z) \in \Xi_{q,s+1,l}$ where

$$
l = n + r
$$

i.e. $Q(p;\beta,z)$ is a matrix form of degree $s+1$ in $p \in \mathbb{R}^q$ for any fixed β and z. Notice that, due to the constraint $p \in \Upsilon_q$ in (4.47), one has

$$Q\left(p;\beta,\frac{1}{\gamma^2}\right) = E(V(p;\beta),p) + \frac{1}{\gamma^2}F(p) \quad \forall p \in \Upsilon_q.$$

Let us write

$$Q(\mathrm{sq}(p);\beta,z) = \Phi\left(R(\beta,z),p^{\{s+1\}},l\right) \tag{4.56}$$

where $R(\beta,z) \in \mathbb{S}^{l\sigma(q,s+1)}$ is any SMR matrix of $Q(\mathrm{sq}(p);\beta,z)$. Observe that $R(\beta,z)$ depends affinely on β and z. Now we are ready to formulate a convex optimization problem whose solution provides an upper bound to $\gamma_s^{H_\infty}$.

Theorem 4.8. *Let $S(\beta)$ be a linear parametrization of $\mathscr{S}_{q,s,n}$ in (4.18), $L(\alpha)$ a linear parametrization of $\mathscr{L}_{q,s+1,l}$ in (1.39), and $R(\beta,z)$ defined according to (4.54)–(4.56). Let*

$$\hat{\gamma}_s^{H_\infty} = \frac{1}{\sqrt{z^*}} \tag{4.57}$$

where z^* is the solution of the EVP

$$z^* = \sup_{z \in \mathbb{R},\ \beta \in \mathbb{R}^{\tau(q,s,n)},\ \alpha \in \mathbb{R}^{\omega(q,s+1,l)}} z \tag{4.58}$$
$$\mathrm{s.t.} \begin{cases} 0 < S(\beta) \\ 0 > R(\beta,z) + L(\alpha). \end{cases}$$

Then, $\hat{\gamma}_s^{H_\infty} \geq \gamma_s^{H_\infty}$.

Proof. Let α and β be such that the LMIs in (4.58) are satisfied. Let us define $V(p) = V(p;\beta)$ with $V(p;\beta)$ defined according to (4.20). From the first LMI in (4.58) and Theorem 4.3 one has that $V(p) > 0$ for all $p \in \Upsilon_q$, and hence the first condition in (4.53) holds. Then, by (4.56) and (4.54) one has

$$\Phi\left(R(\beta,z) + L(\alpha), p^{\{s+1\}}, n\right)$$
$$= \Phi\left(R(\beta,z), p^{\{s+1\}}, n\right)$$
$$= E(V(\mathrm{sq}(p);\beta), \mathrm{sq}(p)) + zF(\mathrm{sq}(p))\left(\sum_{i=1}^q p_i^2\right)^{s-1} + G(\mathrm{sq}(p)).$$

From the second LMI of (4.58) this implies that

$$E(V(\mathrm{sq}(p);\beta), \mathrm{sq}(p)) + zF(\mathrm{sq}(p))\left(\sum_{i=1}^q p_i^2\right)^{s-1} + G(\mathrm{sq}(p)) < 0 \quad \forall p \in \mathbb{R}_0^q.$$

As a consequence, one has from Theorem 1.17 that

$$E(V(p;\beta),p) + zF(p)\left(\sum_{i=1}^q p_i\right)^{s-1} + G(p) < 0 \quad \forall p \in \Upsilon_q$$

which is equivalent to

$$E(V(p;\beta),p)+zF(p)<0 \quad \forall p \in \Upsilon_q.$$

Hence, the second condition in (4.53) is satisfied for $z=\gamma^{-2}$. $\square$

Theorem 4.8 provides an upper bound of $\gamma_s^{H_\infty}$ through the EVP (4.58). The number of scalar variables in the LMI constraint is given by

$$\tau(q,s,n) + \omega(q,s+1,l) + 1.$$

Table 4.2 shows this number for some values of n,s,q,r.

Table 4.2 Number of scalar variables in the EVP (4.58) for some values of n,s,q,r: (a) $q=2$ and $r=1$; (b) $q=3$ and $r=1$

	(a)				(b)			
	$s=0$	$s=1$	$s=2$	$s=3$	$s=0$	$s=1$	$s=2$	$s=3$
$n=2$	7	23	52	94	13	94	349	922
$n=3$	13	44	100	181	25	178	658	1738
$n=4$	21	72	164	297	41	289	1066	2816

The next issue we address is whether there exists a relationship between the upper bound obtained by using HPD-QLFs of degree s and that provided by HPD-QLFs of degree $s+1$. The following result clarifies that, if the condition of Theorem 4.8 is satisfied for s, then it is satisfied also for $s+1$.

Theorem 4.9. *Let $s \geq 0$ be an integer, and $\hat{\gamma}_s^{H_\infty}$ be defined according to (4.57)–(4.58). Then,*

$$\hat{\gamma}_s^{H_\infty} \geq \hat{\gamma}_{s+1}^{H_\infty}. \tag{4.59}$$

Proof. Let us rewrite the LMIs in (4.58) with s replaced by $s+1$ as

$$\begin{cases} 0 < \tilde{S}(\tilde{\beta}) \\ 0 > \tilde{R}(\tilde{\beta},z)+\tilde{L}(\tilde{\alpha}) \end{cases} \tag{4.60}$$

where $\tilde{S}(\tilde{\beta})$ and $\tilde{L}(\tilde{\alpha})$ are linear parametrizations of $\mathscr{S}_{q,s+1,n}$ and $\mathscr{L}_{q,s+2,l}$, respectively; $\tilde{R}(\tilde{\beta},z) \in \mathbb{S}^{l\sigma(q,s+2)}$ is an SMR matrix of $\tilde{Q}(\mathrm{sq}(p);\tilde{\beta},z)$, and $\tilde{Q}(p;\tilde{\beta},z)$ is defined according to (4.54)–(4.55), with s replaced by $s+1$ and $V(p;\beta)$ replaced by

$$\tilde{V}(p;\tilde{\beta}) = \Phi\left(\tilde{S}(\tilde{\beta}), \mathrm{sqr}(p)^{\{s+1\}}, n\right).$$

Let us suppose that there exist z, α, β such that the LMI constraints in (4.58) are satisfied. From the proof of Theorem 4.8 we have that $V(p;\beta) > 0$ and $Q(p;\beta,z) < 0$

for all $p \in \Upsilon_q$. In order to select $\tilde{\beta}$, let us define

$$\tilde{V}(p) = V(p;\beta) \sum_{i=1}^{q} p_i$$

and let us show that $\tilde{V}(\mathrm{sq}(p))$ admits a positive definite SMR matrix. Let T_{s+1} be the matrix defined in (4.28). Then,

$$
\begin{aligned}
\tilde{V}(\mathrm{sq}(p)) &= \left(\sum_{i=1}^{q} p_i^2 \right) \Phi\left(S(\beta), p^{\{s\}}, n \right) \\
&= p'p\, \Phi\left(S(\beta), p^{\{s\}}, n \right) \\
&= \Phi\left(I_q \otimes S(\beta), p \otimes p^{\{s\}}, n \right) \\
&= \Phi\left(I_q \otimes S(\beta), T_{s+1} p^{\{s+1\}}, n \right) \\
&= \Phi\left(\check{S}, p^{\{s+1\}}, n \right)
\end{aligned}
\tag{4.61}
$$

where

$$
\begin{aligned}
\check{S} &= (T_{s+1} \otimes I_n)' \left(I_q \otimes S(\beta) \right) (T_{s+1} \otimes I_n) \\
&= \Phi\left(I_q \otimes S(\beta), T_{s+1}, n \right).
\end{aligned}
$$

Since $S(\beta) > 0$ and T_{s+1} is a matrix with full column rank, it follows that $\check{S} > 0$. Moreover, from (4.61), it is clear that $\check{S} \in \mathscr{S}_{q,s+1,n}$, and hence there exists $\tilde{\beta}$ such that $\tilde{S}(\tilde{\beta}) = \check{S}$.

Now, let us observe that, being $p \in \Upsilon_q$, it is possible to write

$$\tilde{Q}(p;\tilde{\beta},z) = Q(p;\beta,z) \sum_{i=1}^{q} p_i. \tag{4.62}$$

Following the same development as in (4.61), one gets from (4.62)

$$\tilde{Q}(\mathrm{sq}(p);\tilde{\beta},z) = \Phi\left(\check{R}, p^{\{s+2\}}, l \right)$$

where

$$\check{R} = \Phi\left(I_q \otimes (R(\beta,z) + L(\alpha)), T_{s+2}, l \right).$$

Since $R(\beta,z) + L(\alpha) < 0$, one has $\check{R} < 0$. Moreover, being $\check{R}$ and $\tilde{R}(\tilde{\beta},z)$ SMR matrices of the same matrix form, it follows that there exists $\tilde{\alpha}$ such that $\tilde{R}(\tilde{\beta},z) + \check{L}(\tilde{\alpha}) = \check{R}$. Therefore, (4.60) holds, which then yields (4.59). □

The next result shows that it is possible to establish whether the upper bound provided by Theorem 4.8 is tight.

Theorem 4.10. *Suppose that the upper bound $\hat{\gamma}_s^{H_\infty}$ in (4.57)–(4.58) is finite. Then,*

$$\hat{\gamma}_s^{H_\infty} = \gamma^{H_\infty} \tag{4.63}$$

if and only if there exist $\hat{p} \in \Upsilon_q$ and $\hat{x} \in \mathbb{R}_0^l$ satisfying

$$\|H(\cdot,\hat{p})\|_\infty = \hat{\gamma}_s^{H_\infty} \tag{4.64}$$

and

$$\left(\mathrm{sqr}(\hat{p})^{\{s+1\}} \otimes \hat{x}\right) \in \ker(M) \tag{4.65}$$

where

$$M = R(\beta^*, z^*) + L(\alpha^*) \tag{4.66}$$

being z^*, β^*, α^* optimal values of z, β, α in (4.58), and $R(\cdot,\cdot) + L(\cdot)$ the complete SMR matrix in (4.58).

Proof. (Sufficiency) Let us suppose that (4.64)–(4.65) hold. Then, (4.63) holds because $\hat{\gamma}_s^{H_\infty}$ is an upper bound of γ^{H_∞} and because this upper bound is achieved in $\hat{p} \in \Upsilon_q$.

(Necessity) Let us suppose that (4.63) holds. Then, there exists $\hat{p} \in \Upsilon_q$ such that (4.64) holds because Υ_q is a compact set. In order to prove (4.65), let us adopt the same argument as in the proof of Theorem 4.8 and observe that

$$\Phi\left(M, \mathrm{sqr}(\hat{p})^{\{s+1\}}, l\right) = Q(\hat{p}; \beta^*, z^*).$$

Since $M \le 0$ due to (4.58), and since $\hat{p}$ is the point where γ^{H_∞} is attained, one has

$$0 \ge Q(\hat{p}; \beta^*, z^*)$$
$$0 = \det(Q(\hat{p}; \beta^*, z^*)).$$

Indeed, if one assumes by contradiction that $Q(\hat{p}; \beta^*, z^*)$ is nonsingular, then γ^{H_∞} is not the $\mathcal{H}_\infty$ norm of $H(\delta, p)$ due to (4.49). Hence, there exists $\hat{x} \in \mathbb{R}_0^l$ such that

$$\hat{x} \in \ker\left(\Phi\left(M, \mathrm{sqr}(\hat{p})^{\{s+1\}}, l\right)\right).$$

Therefore, one can write

$$0 = \hat{x}' \Phi\left(M, \mathrm{sqr}(\hat{p})^{\{s+1\}}, l\right) \hat{x}$$
$$= \left(\mathrm{sqr}(\hat{p})^{\{s+1\}} \otimes \hat{x}\right)' M \left(\mathrm{sqr}(\hat{p})^{\{s+1\}} \otimes \hat{x}\right)$$

and hence (4.65) holds since M is negative semidefinite.   □

Theorem 4.10 provides a necessary and sufficient condition for establishing tightness of the upper bound $\hat{\gamma}_s^{H_\infty}$. According to this condition, one has to find the vectors $\hat{p} \in \Upsilon_q$ and $\hat{x} \in \mathbb{R}_0^l$ such that (4.65) holds. Then, one has to verify if at least one of these vectors $\hat{p}$ verifies (4.64), for example by using (4.49). The latter step is trivial. In order to perform the first step, let us observe that the vectors $\hat{p}$ and $\hat{x}$ satisfy

$$\mathrm{sqr}(\hat{p})^{\{s+1\}} \otimes \hat{x} = Nv \tag{4.67}$$

where $N \in \mathbb{R}^{l\sigma(q,s+1) \times w}$ is a matrix whose columns form a base of $\ker(M)$, and $v \in \mathbb{R}^w$. The following result suggests a possible way to find vectors $\hat{p}$ and $\hat{x}$ satisfying (4.67).

Theorem 4.11. *The vectors* $\hat{p} \in \Upsilon_q$, $\hat{x} \in \mathbb{R}_0^l$ *and* $v \in \mathbb{R}^w$ *for which* (4.67) *holds, satisfy also*

$$\hat{x}_i \mathrm{sqr}(\hat{p})^{\{s+1\}} = J_i v \quad \forall i = 1, \ldots, l \tag{4.68}$$

and

$$\mathrm{rank}(H) < w. \tag{4.69}$$

The matrix J_i *is defined as*

$$J_i = \left(I_{\sigma(q,s+1)} \otimes e_i' \right) N$$

where e_i *is the i-th column of* I_l. *The matrix* H *is defined as*

$$H = \begin{pmatrix} \sqrt{\hat{p}_b}H_0 - \sqrt{\hat{p}_a}H_1 \\ \sqrt{\hat{p}_b}H_1 - \sqrt{\hat{p}_a}H_2 \\ \vdots \\ \sqrt{\hat{p}_b}H_s - \sqrt{\hat{p}_a}H_{s+1} \end{pmatrix}$$

where $a, b \in [1, q]$ *are integers,* H_i *is given by*

$$H_i = \left(\mathrm{diag}(f_{h(i)}) \otimes I_l \right) N,$$

f_j *is the j-th column of* $I_{\sigma(q,s+1)}$, *and* $h(i)$ *is the position of the monomial* $\hat{p}_a^{s+1-i}\hat{p}_b^i$ *in the power vector* $p^{\{s+1\}}$.

Proof. The condition (4.68) is obtained by multiplying both sides of (4.67) by $I_{\sigma(q,s+1)} \otimes e_i'$. Similarly, the condition (4.69) is obtained by multiplying both sides of (4.67) by $\left(\mathrm{diag}(f_{h(i)}) \otimes I_l \right)$, which provides $(\hat{p}_a^{m+1-i}\hat{p}_b^i)^{1/2}\hat{x} = H_i v$. By eliminating $\hat{x}$ the rank condition on the matrix H follows. $\square$

Theorem 4.11 provides two possible strategies to compute the vectors $\hat{p}$, $\hat{x}$ satisfying (4.65). In particular, since $\hat{x}$ is defined up to a scale factor, one can set $\hat{x}_i = 1$ for some i, and then (4.68) can be used to compute $\hat{p}$ by adopting the approach proposed in Section 1.9 to extract power vector from linear spaces. As an alternative, the condition (4.69) provides a way to compute the ratio between $\hat{p}_a$ and $\hat{p}_b$. In fact, the rank condition $\mathrm{rank}(H) < w$ can be enforced by annihilating the principal minors of H, thus allowing one to compute the entries of $\hat{p}$ by solving polynomial equations in one variable (namely, the square root of the ratio $\hat{p}_a\hat{p}_b^{-1}$).

4.3.2 Parametric Stability Margin

Here we consider the problem of computing the robust parametric stability margin for a time-invariant polytopic system. Let $\gamma \in \mathbb{R}$ and consider the system

$$\begin{cases} \dot{x}(t) = A(p,\gamma)x(t) \\ p \in \Upsilon_q \end{cases} \tag{4.70}$$

with

$$A(p,\gamma) = \sum_{i=1}^{q} p_i A_i(\gamma) \tag{4.71}$$
$$A_i(\gamma) = \bar{A}_0 + \gamma \bar{A}_i, \quad i = 1,\ldots,q$$

where $\bar{A}_0,\ldots,\bar{A}_q$ are given matrices, and $\bar{A}_0$ is Hurwitz.

Definition 4.6 (Parametric Stability Margin). Let us define

$$\gamma^{PA} = \sup \ \{\bar{\gamma} \in \mathbb{R} : \ A(p,\gamma) \text{ is Hurwitz for all } p \in \Upsilon_q, \text{ for all } \gamma \in [0,\bar{\gamma}]\}. \tag{4.72}$$

Then, γ^{PA} is called *parametric stability margin* for the system (4.70)–(4.71).

The robust stability margin can be studied through a one-parameter sequence of robust stability problems, in particular for each fixed value of γ the problem is to establish whether $A(p,\gamma)$ is Hurwitz for all $p \in \Upsilon_q$. We hence define the lower bound of γ^{PA} guaranteed by HPD-QLFs as follows.

Definition 4.7 (s-HPD-QLF Parametric Stability Margin). Let

$$\hat{\gamma}_s^{PA} = \sup \ \{\bar{\gamma} \in \mathbb{R} : \text{ condition (4.23) in Theorem 4.4} \atop \text{ holds for } A(p) = A(p,\gamma), \text{ for all } \gamma \in [0,\bar{\gamma}]\}. \tag{4.73}$$

Then, $\hat{\gamma}_s^{PA}$ is called *s-HPD-QLF parametric stability margin* for the system (4.70)–(4.71).

Clearly, $\hat{\gamma}_s^{PA} \leq \gamma^{PA}$. The following result provides a necessary and sufficient condition for establishing tightness of $\hat{\gamma}_s^{PA}$.

Theorem 4.12. *Let us suppose that $\hat{\gamma}_s^{PA}$ in (4.73) is finite. Let*

$$M = R(\beta^*) + L(\alpha^*) \tag{4.74}$$

where β^*, α^* are optimal values of β, α in (4.41) and the LMIs in (4.41) are constructed for $A(p) = A(p,\hat{\gamma}_s^{PA})$. Then, $\hat{\gamma}_s^{PA} = \gamma^{PA}$ if and only if there exist $\hat{p} \in \Upsilon_q$ and $\hat{x} \in \mathbb{R}_0^n$ such that

$$\left(\mathrm{sqr}(\hat{p})^{\{s+1\}} \otimes \hat{x}\right) \in \ker(M) \tag{4.75}$$

and

$$\max_{\lambda \in \text{spc}(A(\hat{p}, \hat{\gamma}_s^{PA}))} \text{re}(\lambda) = 0. \tag{4.76}$$

Proof. (Necessity) Let us suppose that $\hat{\gamma}_s^{PA} = \gamma^{PA}$. Then, there exists $\hat{p} \in \Upsilon_q$ such that (4.76) holds. Moreover, let us define

$$V(p) = \Phi\left(S(\beta^*), \text{sqr}(p)^{\{s\}}, n\right)$$

and $v(x, p) = x'V(p)x$. Due to (4.76), there exists $\hat{x} \in \mathbb{R}_0^n$ such that $v(\hat{x}, \hat{p}) = 0$. Moreover, let us observe that by construction of M in (4.74) one has

$$\dot{v}(x, p) = x'\Phi\left(M, \text{sqr}(p)^{\{s+1\}}, n\right)x$$

and $M \geq 0$ (in fact, $z^* = 0$ in (4.41) due to the assumption $\hat{\gamma}_s^{PA} = \gamma^{PA}$). Then, (4.75) holds.

(Sufficiency) Direct consequence of $\hat{\gamma}_s^{PA} \leq \gamma^{PA}$ and (4.76). $\qquad\square$

The condition provided by Theorem 4.12 requires to search for vectors $\hat{p} \in \Upsilon_q$ and $\hat{x} \in \mathbb{R}_0^n$ such that (4.75) holds. Notice that (4.75) is analogous to (4.65) in Theorem 4.10. Therefore, one can apply the techniques proposed in Section 4.3.1, according to Theorem 4.11, in order to check tightness of the lower bound $\hat{\gamma}_s^{PA}$.

4.4 Rational Parametric Uncertainty

In this section we want to show that the proposed technique for robustness analysis of time-invariant systems can be applied also when the dependence of $A(p)$ on the uncertain parameter vector p is rational. In particular, consider the system

$$\begin{cases} \dot{x}(t) = A_{rat}(p)x(t) \\ p \in \Upsilon_q \end{cases} \tag{4.77}$$

where $A_{rat}(p) \in \mathbb{R}^{n \times n}$ has the form

$$A_{rat}(p) = \frac{1}{a_2(p)}A_1(p) \tag{4.78}$$

where $A_1(p) \in \mathbb{R}^{n \times n}$ is a matrix polynomial and $a_2(p) \in \mathbb{R}$ is a polynomial. Let us assume that

$$a_2(p) > 0 \quad \forall p \in \Upsilon_q. \tag{4.79}$$

Assumption (4.79) is not restrictive, because if $a_2(p) = 0$ for some $p \in \Upsilon_q$, $A_{rat}(p)$ cannot be Hurwitz unless there are suitable cancellations between the zeros of $A_1(p)$ and $a_2(p)$. Let us express $A_1(p)$ as

$$A_1(p) = \sum_{i=0}^{u} A_{1,i}(p)$$

where u is the degree of $A_1(p)$, and $A_{1,i} \in \Xi^{\sharp}_{q,i,n}$, $i = 1,\ldots,u$. Then, define

$$A_3(p) = \sum_{i=0}^{u} A_{1,i}(p) \left(\sum_{j=0}^{q} p_j \right)^{u-i}. \qquad (4.80)$$

We have that $A_3 \in \Xi^{\sharp}_{q,u,n}$ is a matrix form of degree u. Moreover, the following result holds.

Theorem 4.13. *Let us suppose that (4.79) holds. Then,*

$$A_{rat}(p) \text{ is Hurwitz for all } p \in Y_q$$

if and only if

$$A_3(p) \text{ is Hurwitz for all } p \in Y_q.$$

Proof. For all $p \in Y_q$, one has that

$$A_3(p) = A_1(p) = a_2(p) A_{rat}(p)$$

and the result holds by (4.79). □

As a consequence of Theorem 4.13, one has that the Theorems 4.1, 4.4, 4.5 and 4.6, can be used for studying robust stability of the system (4.77)–(4.79). This simply requires to replace $A(p)$ with $A_3(p)$, and observe that the time derivative of the HPD-QLF $v(x,p) = x'V(p)x$ of degree s in p, along the trajectories of the system (4.77), is now a form of degree $s + u$ rather than $s + 1$ in p, namely

$$\dot{v}(x,p) = x' \, \mathrm{he}\,(V(p)A_3(p))\, x.$$

In particular, the upper bound to the degree of the HPD-QLF in Theorem 4.1 obtained by replacing $A(p)$ with $A_3(p)$ is given by

$$\mu = \frac{1}{2}n(n+1)u - 1.$$

Similar extensions can be formulated for the robust $\mathcal{H}_\infty$ performance of systems with rational parametric dependence.

4.5 Robustness Analysis via Hurwitz Determinants

In this section we derive stability and instability conditions for systems affected by rational parametric uncertainty, by employing a different technique that does not involve Lyapunov functions. Let us consider system (4.77)–(4.79). Let $A_3(p)$ be the matrix form in (4.80) and define the forms

$$z_i(p) = y_i(\mathrm{sq}(p)) \quad \forall i = 1, 2 \tag{4.81}$$

where

$$\begin{aligned} y_1(p) &= \det(-A_3(p)) \\ y_2(p) &= \hbar_{n-1}(A_3(p)) \end{aligned} \tag{4.82}$$

and $\hbar_{n-1}(A_3(p))$ denotes the $(n-1)$-th Hurwitz determinant of $A_3(p)$ (see Appendix A.2 for details). We have the following result.

Theorem 4.14. *Let us consider the system* (4.77)–(4.79), *and let us suppose that*

$$A_3(\bar{p}) \text{ is Hurwitz}, \quad \bar{p} = (1, 0, \ldots, 0)' \tag{4.83}$$

where $A_3(p)$ *is the matrix form in* (4.80). *Then, the system* (4.77)–(4.79) *is robustly stable if and only if*

$$z_i(p) \text{ is positive definite for all } i = 1, 2. \tag{4.84}$$

Proof. From (4.82) it follows that $y_1(p)$ is the product of the eigenvalues of $-A_3(p)$, while $y_2(p)$ satisfies (A.4) in Appendix A.2. Hence, due to the continuity of the eigenvalues of $A_3(p)$ with respect to p and to assumption (4.83), one has that the system (4.77)–(4.79) is robustly stable if and only if

$$y_i(p) > 0 \quad \forall p \in \Upsilon_q, \ \forall i = 1, 2.$$

Therefore, we conclude that the theorem holds by using Theorem 1.17.   □

Let us observe that the assumption (4.83) is clearly not restrictive because, if this assumption does not hold, then $A_3(p)$ cannot be Hurwitz for all $p \in \Upsilon_q$.

It is interesting to observe that one can establish whether (4.84) holds via LMI feasibility tests in a non-conservative way. This is described by the following result.

Theorem 4.15. *Let* $z_i(p)$ *be defined as in* (4.81). *Then, the system* (4.77)–(4.79) *is robustly stable if and only if there exists forms* $h_1(p), h_2(p)$ *such that*

$$\begin{cases} 0 < \lambda(h_i) \\ 0 < \lambda(w_i) \end{cases} \quad \forall i = 1, 2 \tag{4.85}$$

where

$$w_i(p) = h_i(p)z_i(p).$$

Proof. It follows from Theorem 4.14, Theorem 2.1, and Definition 1.7.   □

Let us observe that the condition (4.85) defines a system of two LMIs in the coefficients of $h_1(p), h_2(p)$ and in the variables α required to assess positivity of the SOS index.

The following theorem provides a necessary and sufficient condition to establish whether the system (4.77)–(4.79) is unstable for some admissible values of the uncertain parameter.

Theorem 4.16. *Let $z_i(p)$ be defined as in (4.81). Then, the system (4.77)–(4.79) is not robustly stable if and only if*

$$\exists i \in \{1,2\} : \lambda(z_i) \leq 0 \text{ and } z_i(p) \text{ is SMR-tight.} \tag{4.86}$$

Moreover, $A(p)$ is not Hurwitz for all $p \in \mathcal{Z}_i$ where

$$\mathcal{Z}_i = \left\{ p : p = \tilde{p} \left(\sum_{j=1}^{q} \tilde{p}_j \right)^{-1} \text{ and } \tilde{p} \in \mathrm{mps}(z_i), \ \tilde{p}_j \geq 0 \ \forall j = 1, \dots, q \right\}, \tag{4.87}$$

$\mathrm{mps}(z_i)$ *is the set introduced in Definition 2.5, and i is the integer for which (4.86) holds.*

Proof. It follows from Theorem 4.14 and the definitions of SOS index and SMR-tight form.   □

Example 4.2. Let us consider

$$A(p) = p_1 A_1 + p_2 A_2 + p_3 A_3$$

with

$$A_1 = \begin{pmatrix} -5 & -1 & 4 \\ 0 & -1 & 5 \\ -2 & -2 & -2 \end{pmatrix}, \quad A_2 = \begin{pmatrix} -5 & -1 & 0 \\ 0 & -9 & 5 \\ -6 & -2 & -2 \end{pmatrix}, \quad A_3 = \begin{pmatrix} -5 & 1 & 4 \\ 0 & -1 & -1 \\ 4 & 2 & -2 \end{pmatrix}.$$

The problem consists of establishing whether the system (4.77)–(4.79) is robustly stable or not.

We find that $\lambda(z_1) = -5.0484$ and hence we cannot conclude that the system is robustly stable. Therefore, we attempt to establish instability by using Theorem 4.16. We find that the dimension of the null space in (2.35) is 2. The minimal points set in (2.36) is

$$\text{mps}(z_1) = \left\{ 0.5132 \begin{pmatrix} 1.0000 \\ 0.0000 \\ -1.6726 \end{pmatrix}, 0.5132 \begin{pmatrix} 1.0000 \\ 0.0000 \\ 1.6726 \end{pmatrix} \right\}.$$

Since $\text{mps}(z_1)$ is not empty, we can conclude from Theorem 2.9 that $z_1(p)$ is SMR-tight. Hence, according to Theorem 4.16, we can finally conclude that the system is not robustly stable.

In particular, the set $\mathscr{Z}_1$ in (4.87) is given by

$$\mathscr{Z}_1 = \left\{ \begin{pmatrix} 0.3742 \\ 0.0000 \\ 0.6258 \end{pmatrix} \right\}$$

and the eigenvalues of $A(p)$ for $p \in \mathscr{Z}_1$ are $-6.5571, -1.5136, 0.0707$.

4.6 Discrete-time Systems

In this section, we address robust stability and performance problems for discrete-time systems affected by time-invariant parametric uncertainty.

Let us first consider the system

$$\begin{cases} x(t+1) = A(p)x(t) \\ p \in \Upsilon_q \\ A(p) = \sum_{i=1}^{q} p_i A_i \end{cases} \tag{4.88}$$

where $t \in \mathbb{N}$ is the time, $x(t) \in \mathbb{R}^n$ is the state vector, p is the uncertainty parameter vector, and $A_1, \ldots, A_q \in \mathbb{R}^{n \times n}$ are given matrices. The problem addressed hereafter consists of establishing whether $A(p)$ is Schur for all admissible values of p, according to the following definition (see Appendix A.2 for definition of Schur matrices).

Definition 4.8 (Robust Stability for Discrete-Time Polytopic System). Let us suppose that

$$A \text{ is Schur } \quad \forall A \in \mathscr{A} \tag{4.89}$$

where $\mathscr{A}$ is as in (4.5). Then, $\mathscr{A}$ is said *Schur*, and the system (4.88) is said *robustly stable*.

Therefore, the system (4.88) is robustly stable whenever the matrix $A(p)$ satisfies

$$|\lambda| < 1 \quad \forall \lambda \in \text{spc}(A), \ \forall A \in \mathscr{A}.$$

The following result is the counterpart of Theorem 4.1.

Theorem 4.17. *Let us define*

$$\mu = n(n+1) - 1.$$

The system (4.88) is robustly stable if and only if there exists $V \in \Xi_{q,s,n}$, with $s \leq \mu$, such that

$$\begin{cases} 0 < V(p) \\ 0 > A(p)'V(p)A(p) - V(p) \end{cases} \forall p \in \Upsilon_q.$$

In order to derive the LMI condition for robust stability, let $V(p;\beta)$ be defined by (4.20), and let us introduce

$$Q(p;\beta) = A'(p)V(p;\beta)A(p) - V(p;\beta)\left(\sum_{i=1}^{q} p_i\right)^2. \tag{4.90}$$

We have that $Q(\cdot,\beta) \in \Xi_{q,s+2,n}$ for all β. Moreover, let us observe that

$$Q(p;\beta) = A'(p)V(p;\beta)A(p) - V(p;\beta) \quad \forall p \in \Upsilon_q.$$

Let $R(\beta) \in \mathbb{S}^{n\sigma(q,s+2)}$ be any SMR matrix of $Q(\mathrm{sq}(p);\beta)$, i.e. satisfying

$$Q(\mathrm{sq}(p);\beta) = \Phi\left(R(\beta), p^{\{s+2\}}, n\right). \tag{4.91}$$

Then, Theorems 4.4, 4.5, 4.6 and 4.7 hold for the discrete-time system (4.88) by performing the following modifications:

1. $R(\beta)$ is defined according to (4.91);
2. $L(\alpha)$ is a linear parametrization of $\mathcal{L}_{q,s+2,n}$, with $\alpha \in \mathbb{R}^{\omega(q,s+2,n)}$;
3. the term "Hurwitz" is replaced by "Schur".

HPD-QLFs can be also used to investigate robust $\mathcal{H}_\infty$ performance for discrete-time systems. Indeed, let us consider

$$\begin{cases} x(t+1) = A(p)x(t) + B(p)w(t) \\ \quad y(t) = C(p)x(t) + D(p)w(t) \end{cases} \tag{4.92}$$

where $x \in \mathbb{R}^n$ is the state, $w \in \mathbb{R}^r$ is the input, $y \in \mathbb{R}^g$ is the output, $p \in \Upsilon_q$ is the parametric uncertainty vector, and $A(p) \in \mathbb{R}^{n \times n}$, $B(p) \in \mathbb{R}^{n \times r}$, $C(p) \in \mathbb{R}^{g \times n}$ and $D(p) \in \mathbb{R}^{g \times r}$ are as in (4.48). For any $p \in \Upsilon_q$, the transfer function from w to y is given by

$$H(\delta,p) = C(p)(\delta I_n - A(p))^{-1}B(p) + D(p).$$

For a fixed p, the $\mathcal{H}_\infty$ norm of $H(\delta,p)$ is given by

$$\|H(\cdot,p)\|_\infty = \sup_{\omega \in [0,2\pi]} \|H(e^{j\omega},p)\|_\infty. \tag{4.93}$$

The robust $\mathcal{H}_\infty$ performance of the system (4.92) is defined by γ^{H_∞} in (4.52) with $\|H(\cdot,p)\|_\infty$ as in (4.93). The upper bound of γ^{H_∞} provided by HPD-QLFs of degree s is defined by $\gamma_s^{H_\infty}$ in (4.53) with $E(V(p),p)$ given by

$$E(V,p) = \begin{pmatrix} A(p)'VA(p) - V & A(p)'VB(p) \\ \star & B(p)'VB(p) - I_r \end{pmatrix}. \tag{4.94}$$

Let $V(p;\beta)$ be defined by (4.20), and let us introduce

$$Q(p;\beta,z) = E(V(p;\beta),p) + z \left(\sum_{i=1}^{q} p_i \right)^s F(p) + G(p) \tag{4.95}$$

where $E(V(p;\beta),p)$ is defined by (4.94), $F(p)$ is as in (4.51), and

$$G(p) = \begin{pmatrix} 0_{n \times n} & 0_{n \times r} \\ \star & \left(1 - \left(\sum_{i=1}^{q} p_i\right)^{s+2}\right) I_r \end{pmatrix}. \tag{4.96}$$

We have that Theorems 4.8, 4.9 and 4.10 hold for the discrete-time system (4.92) by performing the following changes:

1. $R(\beta,z)$, is an SMR matrix of $Q(\mathrm{sq}(p);\beta,z)$, defined according to (4.95);
2. $L(\alpha)$ is a linear parametrization of $\mathcal{L}_{q,s+2,l}$, with $\alpha \in \mathbb{R}^{\omega(q,s+2,l)}$.

Lastly, we address the case of discrete-time systems with rational parametric dependence. Let us consider

$$x(t+1) = A_{rat}(p)x(t) \tag{4.97}$$

where $A_{rat}(p)$ is as in (4.78). Let us define the forms

$$z_i(p) = y_i(\mathrm{sq}(p)) \quad \forall i = 1,2,3 \tag{4.98}$$

where

$$\begin{aligned} y_1(p) &= \det(I_n - A_3(p)) \\ y_2(p) &= \det(I_n + A_3(p)) \\ y_3(p) &= \det(I_d - Q(A_3(p))), \end{aligned} \tag{4.99}$$

$d = \frac{1}{2}n(n-1)$, and $Q(A(p))$ is the matrix defined in Appendix A.2. We have that Theorems 4.13, 4.14, 4.15 and 4.16 hold for the discrete-time system (4.97) by performing the following changes:

1. the term "Hurwitz" is replaced by "Schur";
2. the index i goes from 1 to 3;
3. $z_i(p)$ is defined as in (4.98)–(4.99).

4.7 Examples

In this section, numerical examples are presented to illustrate the proposed techniques for robustness analysis of time-invariant polytopic systems.

4.7.1 Example HPD-QLF-1

Let us start with a very simple example, in order to show how the LMIs involved in robust stability analysis are generated. Let us consider system (4.70)–(4.71) with $q = 2, n = 2$ and

$$\bar{A}_0 = \begin{pmatrix} 0 & 1 \\ -2 & -2 \end{pmatrix}, \quad \bar{A}_1 = \begin{pmatrix} -1 & 0 \\ 0 & 5 \end{pmatrix}, \quad \bar{A}_2 = \begin{pmatrix} 1 & 0 \\ 0 & -5 \end{pmatrix}.$$

We want to compute the parametric stability margin (4.72), and hence we calculate its lower bound in (4.73) guaranteed by an HPD-QLF of degree s.

Let us first consider $s = 0$, which means that a common Lyapunov function is sought for all the matrices of the polytope $\mathscr{A}$ (in other words, the Lyapunov function does not depend on the uncertain parameter). The LMI feasibility test (4.23) involves the matrices

$$S(\beta) = \begin{pmatrix} \beta_1 & \beta_2 \\ \star & \beta_3 \end{pmatrix}, \quad R(\beta) = \begin{pmatrix} r_1 & r_2 & 0 & 0 \\ \star & r_3 & 0 & 0 \\ \star & \star & r_4 & r_5 \\ \star & \star & \star & r_6 \end{pmatrix}, \quad L(\alpha) = \begin{pmatrix} 0 & 0 & 0 & -\alpha_1 \\ \star & 0 & \alpha_1 & 0 \\ \star & \star & 0 & 0 \\ \star & \star & \star & 0 \end{pmatrix}$$

where

$$\begin{aligned}
r_1 &= 2\gamma\beta_1 + 4\beta_2, & r_2 &= -\beta_1 - 2(2\gamma - 1)\beta_2 + 2\beta_3 \\
r_3 &= -2\beta_2 - 2(5\gamma - 2)\beta_3, & r_4 &= -2\gamma\beta_1 + 4\beta_2 \\
r_5 &= -\beta_1 + 2(2\gamma + 1)\beta_2 + 2\beta_3, & r_6 &= 2 - \beta_2 + 2(5\gamma + 2)\beta_3.
\end{aligned}$$

Hence, the number of scalar variables is 4, and the 0-HPD-QLF parametric stability margin provided by (4.73) is $\hat{\gamma}_0^{PA} = 0.3209$.

Let us consider now the case $s = 1$, which corresponds to choosing a linearly parameter-dependent quadratic Lyapunov function. In this case, the matrices involved in (4.23) are

$$S(\beta) = \begin{pmatrix} \beta_1 & \beta_2 & 0 & -\beta_7 \\ \star & \beta_3 & \beta_7 & 0 \\ \star & \star & \beta_4 & \beta_5 \\ \star & \star & \star & \beta_6 \end{pmatrix}, \quad R(\beta) = \begin{pmatrix} r_1 & r_2 & 0 & 0 & 0 & 0 \\ \star & r_3 & 0 & 0 & 0 & 0 \\ \star & \star & r_7 & r_8 & 0 & 0 \\ \star & \star & \star & r_9 & 0 & 0 \\ \star & \star & \star & \star & r_4 & r_5 \\ \star & \star & \star & \star & \star & r_6 \end{pmatrix},$$

$$L(\alpha) = \begin{pmatrix} 0 & 0 & 0 & -\alpha_1 & \alpha_3 & -\alpha_2 - \alpha_4 \\ \star & 0 & \alpha_1 & 0 & \alpha_2 & -\alpha_5 \\ \star & \star & 2\alpha_3 & \alpha_4 & 0 & -\alpha_6 \\ \star & \star & \star & 2\alpha_5 & \alpha_6 & 0 \\ \star & \star & \star & \star & 0 & 0 \\ \star & \star & \star & \star & \star & 0 \end{pmatrix},$$

where $r_1, \ldots, r_6$ are as in the case $s = 0$, while

$$r_7 = -2\gamma\beta_1 + 4\beta_2 + 2\gamma\beta_4 + 4\beta_5$$
$$r_8 = -\beta_1 + 2(2\gamma + 1)\beta_2 + 2\beta_3 - \beta_4 - 2(2\gamma - 1)\beta_5 + 2\beta_6$$
$$r_9 = -2\beta_2 + 2(5\gamma + 2)\beta_3 - 2\beta_5 - 2(5\gamma - 2)\beta_6.$$

The number of scalar variables is hence increased to 13, and the 1-HPD-QLF parametric stability margin provided by (4.73) is $\hat{\gamma}_1^{PA} = 0.4633$. From Theorem 4.12 we find that this lower bound is tight, indeed

$$\ker(M) = \mathrm{img} \begin{pmatrix} 0.2347 & -0.0000 \\ 0.9683 & -0.0000 \\ -0.0000 & -0.0000 \\ -0.0000 & -0.0000 \\ 0.0367 & 0.9073 \\ 0.0771 & -0.4204 \end{pmatrix}$$

and conditions (4.76)–(4.75) hold with

$$\hat{p} = \begin{pmatrix} 0.0000 \\ 1.0000 \end{pmatrix}, \quad \hat{x} = \begin{pmatrix} 0.9073 \\ -0.4204 \end{pmatrix}$$

in particular

$$\mathrm{spc}\left(A(\hat{p}, \hat{\gamma}_1^{PA})\right) = \{0.0000, -3.8533\}.$$

Therefore,

$$\gamma^{PA} = \hat{\gamma}_0^{PA} = 0.4633.$$

It is worth observing that, for this example, γ^{PA} can be computed analytically, and one finds $\gamma^{PA} = (\sqrt{11} - 1)/5 = 0.4633$.

4.7.2 Example HPD-QLF-2

Consider the problem of computing the robust parametric margin γ^{PA} defined in (4.72), for the system (4.70)–(4.71) with $q = 3$, $n = 4$ and

$$\bar{A}_0 = \begin{pmatrix} -2.4 & -0.6 & -1.7 & 3.1 \\ 0.7 & -2.1 & -2.6 & -3.6 \\ 0.5 & 2.4 & -5.0 & -1.6 \\ -0.6 & 2.9 & -2.0 & -0.6 \end{pmatrix}, \quad \bar{A}_1 = \begin{pmatrix} 1.1 & -0.6 & -0.3 & -0.1 \\ -0.8 & 0.2 & -1.1 & 2.8 \\ -1.9 & 0.8 & -1.1 & 2.0 \\ -2.4 & -3.1 & -3.7 & -0.1 \end{pmatrix},$$

$$\bar{A}_2 = \begin{pmatrix} 0.9 & 3.4 & 1.7 & 1.5 \\ -3.4 & -1.4 & 1.3 & 1.4 \\ 1.1 & 2.0 & -1.5 & -3.4 \\ -0.4 & 0.5 & 2.3 & 1.5 \end{pmatrix}, \quad \bar{A}_3 = \begin{pmatrix} -1.0 & -1.4 & -0.7 & -0.7 \\ 2.1 & 0.6 & -0.1 & -2.1 \\ 0.4 & -1.4 & 1.3 & 0.7 \\ 1.5 & 0.9 & 0.4 & -0.5 \end{pmatrix}.$$

The lower bounds in (4.73) are:

$$\hat{\gamma}_0^{PA} = 1.0191, \quad \hat{\gamma}_1^{PA} = 1.9680, \quad \hat{\gamma}_2^{PA} = 2.2238.$$

The number of scalar variables in the LMI feasibility test (4.23) for these lower bounds is equal to 29, 199 and 751, respectively (see also Table 4.1).

From Theorem 4.12 we find that $\hat{\gamma}_2^{PA}$ is tight, i.e. $\hat{\gamma}_2^{PA} = \gamma^{PA}$, indeed conditions (4.76)–(4.75) hold with

$$\hat{p} = \begin{pmatrix} 0.0000 \\ 1.0000 \\ 0.0000 \end{pmatrix}, \quad \hat{x} = \begin{pmatrix} 0.7469 \\ -0.3988 \\ -0.2786 \\ 0.4534 \end{pmatrix}$$

in particular

$$\text{spc}\left(A(\hat{p}, \hat{\gamma}_2^{PA})\right) = \{0.0000 \pm 7.5856i, -5.6059 \pm 1.2008i\}.$$

4.7.3 Example HPD-QLF-3

In this example we want to show how HPD-QLFs can be used to detect instability, by using Theorem 4.7. Let us consider the system (4.1)–(4.3) with $q = 3$, $n = 3$ and

$$A_1 = \begin{pmatrix} -2.5 & 0.0 & -0.5 \\ 1.0 & -3.0 & 1.0 \\ 1.0 & 1.0 & -1.0 \end{pmatrix}, \quad A_2 = \begin{pmatrix} -0.5 & 0.0 & 1.0 \\ 0.0 & -3.0 & 2.0 \\ 0.0 & 0.5 & -1.0 \end{pmatrix},$$

$$A_3 = \begin{pmatrix} -0.5 & 0.0 & -0.5 \\ 0.0 & -3.0 & 2.0 \\ 1.0 & 2.0 & -1.0 \end{pmatrix}.$$

Let us use an HPD-QLF with $s = 1$. We find that z^* in (4.41) satisfies $z^* \leq 0$, and that the set $\mathscr{P}_M$ in (4.46) contains the vector $\hat{p} = (0.0236, 0.4710, 0.5054)'$. It turns out that $A(\hat{p})$ is not Hurwitz, indeed one has

$$\mathrm{spc}(A(\hat{p})) = \{0.0245, -0.6911, -3.8806\}.$$

Therefore, the system is not robustly stable.

4.7.4 Example HPD-QLF-4

Here we consider the problem of studying robust $\mathscr{H}_\infty$ performance. Let us consider the system (4.47)–(4.48) with $q = 2$ and

$$A_1 = \hat{A}_0 + \kappa \hat{A}_1, \quad A_2 = \hat{A}_0 - \kappa \hat{A}_1$$

where $\kappa \in \mathbb{R}$ and

$$\hat{A}_0 = \begin{pmatrix} -2.0 & 1.0 & -1.0 \\ 2.5 & -3.0 & 0.5 \\ -1.0 & 1.0 & -3.5 \end{pmatrix}, \quad \hat{A}_1 = \begin{pmatrix} -0.7 & -0.5 & -2.0 \\ -0.8 & 0.0 & 0.0 \\ 1.5 & 2.0 & 2.4 \end{pmatrix}.$$

The matrices B_i, C_i and D_i are given by

$$B_i = \begin{pmatrix} 1 \\ 0 \\ 0 \end{pmatrix}, \quad C_i = \begin{pmatrix} 0 \\ 0 \\ 1 \end{pmatrix}', \quad D_i = 0$$

for $i = 1, 2$. Table 4.3 shows the upper bounds to the robust $\mathscr{H}_\infty$ performance γ^{H_∞}, for some values of κ (the semi-length of the segment of matrices $A(p)$), provided by Theorem 4.8 with $s = 1$ (linear dependence) and $s = 2$ (quadratic dependence). The tightness of each computed upper bound is investigated through Theorem 4.10. As it can be seen from Table 4.3, the upper bound $\hat{\gamma}_2^{H_\infty}$ always turns out to be tight; in particular, the value of $\hat{p}$ satisfying the condition of Theorem 4.10 is shown. Notice that the maximum value of κ for which $A(p)$ is Hurwitz is $\kappa = 3.552$, which has been computed by exploiting Theorems 4.4 and 4.12.

Table 4.3 Example HPD-QLF-4: upper bounds $\hat{\gamma}_s^{H_\infty}$ for some values of s, κ

κ	$\hat{\gamma}_1^{H_\infty}$	tight	$\hat{\gamma}_2^{H_\infty}$	tight	$\hat{p}$ for $\hat{\gamma}_2^{H_\infty}$
1.6	1.5673	yes	1.5673	yes	$(0.1403, 0.8597)'$
1.8	1.5673	yes	1.5673	yes	$(0.1803, 0.8197)'$
2.0	1.6160	no	1.5673	yes	$(0.2122, 0.7878)'$
2.2	2.2608	yes	2.2608	yes	$(1.0000, 0.0000)'$
2.4	3.5001	yes	3.5001	yes	$(1.0000, 0.0000)'$
2.6	5.2545	no	5.2320	yes	$(1.0000, 0.0000)'$
2.8	7.9907	no	6.2009	yes	$(0.9979, 0.0021)'$
3.0	15.5001	no	6.2009	yes	$(0.9647, 0.0353)'$
3.2	398.5428	no	6.2009	yes	$(0.9356, 0.0644)'$
3.4	∞	no	6.2009	yes	$(0.9100, 0.0900)'$
3.5	∞	no	6.2009	yes	$(0.8984, 0.1016)'$

4.8 Notes and References

Robustness analysis of uncertain systems affected by time-invariant structured parametric uncertainty has been the subject of an intense research activity in the last four decades, see for instance the books [139, 3, 5] and references therein. Unfortunately, while for robust stability of polytopes of polynomials results have been found which lead to a significant reduction in the computational complexity, this is not the case for polytopes of matrices [54]. Therefore, sufficient conditions for robust stability have been derived by exploiting Lyapunov stability theory.

Common quadratic Lyapunov functions have been largely employed, see e.g. [79, 140, 112], but they are known to be conservative. In order to reduce conservatism, parameter-dependent quadratic Lyapunov functions have been proposed, see [3]. Among the very large number of contributions, we recall [56, 99, 58, 108, 120, 90, 59] for Lyapunov functions with linear dependence on the uncertain parameters. Polynomial dependence on the uncertainty has been considered in [10, 11] for systems with parameters constrained in a hypercube, and in [149, 124, 63] for systems with single parameter. Convex relaxations for Lyapunov functions with polynomial parameter dependence have been devised by employing SOS-based techniques [77, 126, 127, 128, 88], matrix dilation approaches [109, 125, 100], and moments theory [60].

HPD-QLFs have been introduced in [42]. Most results reported in this chapter have been presented in [40, 41, 24, 26]. Further results on robustness analysis via HPD-QLFs can be found in [12, 103]. LMI-based robustness conditions based on Lyapunov functions which are polynomial functions of the uncertain system matrices, have been proposed in [102]. The conditions for robust stability and instability based on Hurwitz determinants have been proposed in [23, 27].

Chapter 5
Robustness with Bounded-rate Time-varying Uncertainty

This chapter addresses robustness analysis of polytopic systems affected by time-varying uncertainties with known bounds on their variation rate. The analysis is conducted by introducing the class of HPD-HLFs, i.e. Lyapunov functions that are forms in both the state and the uncertain parameters, which includes the classes of HPLFs and HPD-QLFs as special cases. It is shown that the construction of HPD-HLFs for assessing robust stability as well as computing robust stability margins can be tackled via convex optimizations constrained by LMIs through the use of the SMR of matrix forms.

5.1 Polytopic Systems with Bounded-rate Time-varying Uncertainty

Let us start by considering linear systems affected by linear dependent time-varying parametric uncertainty with bounds on the variation rate.

Definition 5.1 (Bounded-rate Time-varying Polytopic System). Let us consider the continuous-time system described by

$$\dot{x}(t) = A(p(t))x(t) \tag{5.1}$$

where $x(t) \in \mathbb{R}^n$ is the state vector, $A(p(t)) \in \mathbb{R}^{n \times n}$ is a linear function expressed as

$$A(p(t)) = \sum_{i=1}^{q} p_i(t) A_i$$

and $A_1, \ldots, A_q \in \mathbb{R}^{n \times n}$ are given matrices. The uncertain parameter vector $p(t) \in \mathbb{R}^q$ is supposed to be a continuously differentiable function of time, satisfying the constraints

$$\begin{cases} p(t) \in \Upsilon_q \\ \dot{p}(t) \in \mathscr{D} \end{cases} \tag{5.2}$$

G. Chesi et al.: Homogeneous Polynomial Forms, LNCIS 390, pp. 133–153.
springerlink.com   © Springer-Verlag Berlin Heidelberg 2009

where Υ_q is the simplex in (1.68), $\dot{p}(t) = \frac{dp(t)}{dt}$ is the time derivative of $p(t)$, $\mathscr{D}$ is a polytope given by

$$\mathscr{D} = \mathrm{co}\left\{d^{(1)},\ldots,d^{(h)}\right\} \tag{5.3}$$

with $d^{(1)},\ldots,d^{(h)} \in \mathbb{R}^q$ given vectors such that

$$\begin{cases} 0_q \in \mathscr{D} \\ \sum_{i=1}^{q} d_i^{(j)} = 0 \;\; \forall j = 1,\ldots,h. \end{cases} \tag{5.4}$$

Then, the system (5.1)–(5.4) is called *bounded-rate time-varying polytopic system*.

Let us observe that the second constraint in (5.2) imposes bounds on the variation rate of the uncertain vector $p(t)$. These bounds are expressed via the set $\mathscr{D}$ that, for reasons that will become clear in the sequel, is chosen as a polytope.

The first condition in (5.4) is included in order to consider arbitrarily slow variation rates. Instead, the second condition in (5.4) is necessary in order to ensure that the system (5.1) is well posed. Indeed, since $p(t)$ has to belong to Υ_q according to the first constraint in (5.2), one has that

$$\sum_{i=1}^{q} p_i(t) = 1$$

which implies that

$$\sum_{i=1}^{q} \dot{p}_i(t) = 0.$$

Moreover, since $\dot{p}(t)$ has to belong to $\mathscr{D}$ according to the second constraint in (5.2), one has that

$$\dot{p}(t) = \sum_{j=1}^{h} c_j(t) d^{(j)}$$

where $c_1(t),\ldots,c_h(t) \in \mathbb{R}$ are such that

$$\begin{cases} \sum_{j=1}^{h} c_j(t) = 1 \\ c_j(t) \geq 0 \;\; \forall j = 1,\ldots,h. \end{cases}$$

Hence:

$$\sum_{i=1}^{q} \dot{p}_i(t) = \sum_{j=1}^{h} c_j(t) \sum_{i=1}^{q} d_i^{(j)} = 0$$

which is ensured by the second condition in (5.4).

It is useful to observe that the model (5.1)–(5.4) contains as special cases the models introduced in Chapters 3 and 4. Indeed:

1. the model (4.1)–(4.3) for time-invariant polytopic systems is obtained by selecting $h = 1$ and $d^{(1)} = 0_q$ in (5.1)–(5.4);
2. the model (3.1)–(3.4) for time-varying polytopic system is obtained by selecting q equal to the number of vertices r of the polytope $\mathscr{P}$ in (3.4), and $A_1, \ldots, A_q$ equal to the matrices $A(p^{(i)}), \ldots, A(p^{(r)})$ in (3.2)–(3.4). In this case, $\mathscr{D}$ is given by $\mathbb{R}^q$.

A fundamental problem for bounded-rate time-varying polytopic systems is to establish whether the uncertain system is robustly stable with respect to the admissible uncertainty.

Definition 5.2 (Robust Stability for Bounded-rate Time-varying Polytopic System). The system (5.1)–(5.4) is said *robustly stable* if the following conditions hold:

1. $\forall \varepsilon > 0 \; \exists \delta > 0 : \; \|x(0)\| < \delta \Rightarrow \|x(t)\| \leq \varepsilon \;\; \forall t \geq 0, \; \forall p(t) \in \mathscr{P}, \; \forall \dot{p}(t) \in \mathscr{D}$;
2. $\lim_{t \to \infty} x(t) = 0_n \;\; \forall x(0) \in \mathbb{R}^n, \; \forall p(t) \in \mathscr{P}, \; \forall \dot{p}(t) \in \mathscr{D}$.

Hence, the system (5.1)–(5.4) is robustly stable whenever its origin is a globally asymptotically stable equilibrium point for all admissible parametric uncertainties. In the sequel, the dependence of matrices and vectors on time t will be omitted for conciseness unless specified otherwise.

In order to investigate robust stability of the system (5.1)–(5.4), we introduce the following class of parameter-dependent Lyapunov functions.

Definition 5.3 (HPD-HLF). Let $v : \mathbb{R}^n \times \varUpsilon_q \to \mathbb{R}$ be a function satisfying

$$\begin{cases} v(\cdot, p) \in \varXi_{n,2m} \;\; \forall p \in \varUpsilon_q \\ v(x, \cdot) \in \varXi_{q,s} \;\; \forall x \in \mathbb{R}^n \\ v(x, p) > 0 \;\; \forall x \in \mathbb{R}_0^n, \; \forall p \in \varUpsilon_q \\ \dot{v}(x, p) < 0 \;\; \forall x \in \mathbb{R}_0^n, \; \forall p \in \varUpsilon_q, \; \forall \dot{p} \in \mathscr{D} \end{cases} \quad (5.5)$$

where

$$\dot{v}(x, p) = \left. \frac{dv(x, p)}{dt} \right|_{\dot{x} = A(p)x}. \quad (5.6)$$

Then, $v(x, p)$ is said an HPD-HLF of degree $2m$ in $x \in \mathbb{R}^n$ and degree s in $p \in \mathbb{R}^q$ for the system (5.1)–(5.4).

Hence, HPD-HLFs are forms proving stability of the origin of the system (5.1)–(5.4), for all admissible values of the uncertain vector and its time derivative.

A HPD-HLF can be written as

$$v(x, p) = \sum_{\substack{i \in \mathbb{N}^q \,:\, \sum_{k=1}^q i_k = s \\ j \in \mathbb{N}^n \,:\, \sum_{k=1}^n j_k = 2m}} a_{i,j} p^i x^j \quad (5.7)$$

where $a_{i,j} \in \mathbb{R}$ are coefficients. Such a class of Lyapunov functions can be seen as a generalization of the HPLFs introduced in Chapter 3 for time-varying uncertainty, and the HPD-QLFs introduced in Chapter 4 for time-invariant uncertainty. Indeed:

1. HPLFs are recovered from HPD-HLFs with the choice $s = 0$;
2. HPD-QLFs are recovered from HPD-HLFs with the choice $m = 1$.

Let us also observe that affine parameter-dependent quadratic Lyapunov functions are singled out for $s = 1$ and $m = 1$.

It is worth noticing that the choice of Lyapunov functions $v(x, p)$ homogeneous in the uncertain vector p is not conservative with respect to a more general polynomial dependence. This is due to the fact that in the system (5.1)–(5.4), the uncertain vector p belongs to the simplex. Indeed, the following result holds.

Lemma 5.1. *Let $v_1(x, p)$ be a form of degree $2m$ in x for any fixed p, and a polynomial of degree s in p for any fixed x, and let us suppose that $v_1(x, p)$ is a Lyapunov function for the system (5.1)–(5.4), i.e.:*

1. $v_1(0_n; p) = 0$ and $v_1(x, p) > 0$ $\forall x \in \mathbb{R}_0^n$, $\forall p \in \Upsilon_q$;
2. $\dot{v}_1(x, p) < 0$ $\forall x \in \mathbb{R}_0^n$, $\forall p \in \Upsilon_q$, $\forall \dot{p} \in \mathscr{D}$.

Then, there exists an HPD-HLF for the system (5.1)–(5.4).

Proof. Let us write $v_1(x, p)$ as

$$v_1(x, p) = \sum_{\substack{i \in \mathbb{N}^q \, : \, \sum_{k=1}^q i_k \leq s \\ j \in \mathbb{N}^n \, : \, \sum_{k=1}^n j_k = 2m}} a_{1,i,j} p^i x^j$$

for some coefficients $a_{1,i,j} \in \mathbb{R}$. Then, let us define the function

$$v(x, p) = \sum_{\substack{i \in \mathbb{N}^q \, : \, \sum_{k=1}^q i_k \leq s \\ j \in \mathbb{N}^n \, : \, \sum_{k=1}^n j_k = 2m}} a_{1,i,j} \left(\sum_{k=1}^q p_k \right)^{s - \sum_{k=1}^q i_k} p^i x^j.$$

It follows that $v(x, p) = v_1(x, p)$ for all $p \in \Upsilon_q$, and hence $v(x, p)$ is an HPD-HLF for the system (5.1)–(5.4). $\qquad\square$

5.2 Robust Stability

This section investigates robust stability of the system (5.1)–(5.4) by using HPD-HLFs and the results derived in Chapters 1–2.

5.2.1 *Parametrization of HPD-HLFs*

The aim is to find an HPD-HLF as in (5.7) for chosen degrees $2m$ and s. The conditions to be satisfied are the positive definiteness of $v(x, p)$ and negative definiteness of $\dot{v}(x, p)$ with respect to x, for all $p \in Y_q$ and $\dot{p} \in \mathscr{D}$. In this respect, a parametrization of functions fulfilling this property is provided next.

Theorem 5.1. *Let* $v(x, p)$ *be a form in* x *for any fixed* p *and a form in* p *for any fixed* x. *Then,* $v(x, p)$ *is an HPD-HLF for the system* (5.1)–(5.4) *if and only if*

$$\begin{cases} v(x, \mathrm{sq}(p)) > 0 \\ \dot{v}(x, \mathrm{sq}(p)) < 0 \end{cases} \quad \forall x \in \mathbb{R}_0^n, \ \forall p \in \mathbb{R}_0^q, \ \forall \dot{p} \in \mathscr{D}. \tag{5.8}$$

Proof. By Theorem 1.17, it follows that (5.8) is equivalent to the conditions in (5.5), and hence the theorem holds. $\qquad\square$

Theorem 5.1 states that one can get rid of the constraint $p \in Y_q$ in the problem of investigating positivity and negativity of $v(x, p)$ and $\dot{v}(x, p)$, respectively, by considering these functions evaluated in $\mathrm{sq}(p)$ rather than in p.

Hence, let us first consider the first condition in (5.8) (the second condition will be addressed in the next subsection). We want to find a suitable representation for the function $v(x, \mathrm{sq}(p))$, which is a form of degree $2m$ in x for any fixed p, and a form of degree $2s$ in p for any fixed x. To this aim, we can exploit the SMR. Indeed, let us introduce the notation

$$\Psi\left(S, p^{\{s\}}, x^{\{m\}}\right) = \left(p^{\{s\}} \otimes x^{\{m\}}\right)' S\left(p^{\{s\}} \otimes x^{\{m\}}\right) \tag{5.9}$$

for some $S \in \mathbb{S}^{\sigma(q,s)\sigma(n,m)}$. It follows that $v(x, \mathrm{sq}(p))$ can be represented as

$$v(x, \mathrm{sq}(p)) = \Psi\left(S, p^{\{s\}}, x^{\{m\}}\right) \tag{5.10}$$

for a suitable $S \in \mathbb{S}^{\sigma(q,s)\sigma(n,m)}$. Such a matrix S is said an SMR matrix of $v(x, \mathrm{sq}(p))$ with respect to the vector $p^{\{s\}} \otimes x^{\{m\}}$.

Clearly, the matrix S in (5.10) must have a special structure, in particular it has to belong to the set

$$\mathscr{S}_{q,s,n,m} = \left\{S \in \mathbb{S}^{\sigma(q,s)\sigma(n,m)} : \Psi\left(S, p^{\{s\}}, x^{\{m\}}\right) \text{ does not contain} \right.$$
$$\left. \text{monomials } p_1^{i_1} \dots p_q^{i_q} x_1^{j_1} \dots x_n^{j_n} \text{ with any odd } i_1, \dots, i_q\right\}. \tag{5.11}$$

In fact, it is straightforward to verify that (5.10) holds for some symmetric matrix S if and only if such a matrix S belongs to the set $\mathscr{S}_{q,s,n,m}$. Hence, due to (5.10), one has that

$$S > 0 \implies v(x, \mathrm{sq}(p)) > 0 \quad \forall x \in \mathbb{R}_0^n, \ \forall p \in \mathbb{R}_0^q.$$

Moreover, due to Theorem 5.1, one has that

$$S > 0 \implies v(x, p) > 0 \quad \forall x \in \mathbb{R}_0^n, \ \forall p \in Y_q.$$

In order to increase the degrees of freedom in the selection of $v(x,p)$, one can exploit the fact that the matrix S in (5.10) is not unique. Indeed, the following result provides a characterization of the set $\mathscr{S}_{q,s,n,m}$.

Theorem 5.2. *The set $\mathscr{S}_{q,s,n,m}$ is a linear space of dimension*

$$\tau(q,s,n,m) = \frac{1}{2}\sigma(q,s)\sigma(n,m)\left(\sigma(q,s)\sigma(n,m)+1\right) \\ -\left(\sigma(q,2s)-\sigma(q,s)\right)\sigma(n,2m). \tag{5.12}$$

Proof. If $S_1, S_2 \in \mathscr{S}_{q,s,n,m}$, then $a_1 S_1 + a_2 S_2 \in \mathscr{S}_{q,s,n,m}$ for all $a_1, a_2 \in \mathbb{R}$, hence implying that $\mathscr{S}_{q,s,n,m}$ is a linear space. Now, let us consider (5.12), and let us observe that $\frac{1}{2}\sigma(q,s)\sigma(n,m)\left(\sigma(q,s)\sigma(n,m)+1\right)$ is the number of distinct entries of a symmetric matrix of size $\sigma(q,s)\sigma(n,m)$, whereas $\left(\sigma(q,2s)-\sigma(q,s)\right)\sigma(n,2m)$ is the total number of monomials in $\Psi\left(s,p^{\{s\}},x^{\{m\}}\right)$ containing at least one odd power of the variables $p_1, \ldots, p_q$. The constraints obtained by annihilating these monomials are linear and independent similarly to the proof of Theorem 1.2. Therefore, the dimension of $\mathscr{S}_{q,s,n,m}$ is given by $\tau(q,s,n,m)$. $\square$

Table 5.1 reports $\tau(q,s,n,m)$ for some values of q,s,n,m.

Table 5.1 $\tau(q,s,n,m)$ for some values of q,s,n,m: (a) $q=2$ and $n=2$; (b) $q=2$ and $n=3$

(a)					(b)			
	$s=1$	$s=2$	$s=3$	$s=4$		$s=1$	$s=2$	$s=3$	$s=4$
$m=1$	7	15	27	43	$m=1$	15	33	60	96
$m=2$	16	35	63	100	$m=2$	63	141	255	405
$m=3$	29	64	115	182	$m=3$	182	409	736	1163
$m=4$	46	102	183	289	$m=4$	420	945	1695	2670

Now, we ask how to generate HPD-HLF candidates: the following result, which stems from the definition of $\mathscr{S}_{q,s,n,m}$ and Theorem 5.1, provides an answer to this question.

Theorem 5.3. *Let $S(\beta)$ be a linear parametrization of $\mathscr{S}_{q,s,n,m}$ in (5.11), with $\beta \in \mathbb{R}^{\tau(q,s,n,m)}$. Let $v(x,p;\beta)$ be the function defined according to*

$$v(x,\mathrm{sq}(p);\beta) = \Psi\left(S(\beta),p^{\{s\}},x^{\{m\}}\right). \tag{5.13}$$

Then,

$$v(\cdot;p) \in \Xi_{n,2m} \text{ and } v(x;\cdot) \in \Xi_{q,s} \iff \exists\beta : v(x,p) = v(x,p;\beta).$$

Moreover,

$$\exists \beta : S(\beta) > 0 \;\Rightarrow\; v(x, p; \beta) > 0 \quad \forall x \in \mathbb{R}_0^n, \, \forall p \in Y_q.$$

5.2.2 Robust Stability Condition

In the following, a sufficient condition for the solution of the robust stability problem is provided. To this purpose, let us consider $v(x, p; \beta)$ defined according to (5.13) and write it as

$$
\begin{aligned}
\dot{v}(x, p; \beta) &= \left. \frac{dv(x, p; \beta)}{dt} \right|_{\dot{x} = A(p)x} \\
&= \frac{\partial v(x, p; \beta)}{\partial x} A(p)x + \frac{\partial v(x, p; \beta)}{\partial p} \dot{p}.
\end{aligned}
\tag{5.14}
$$

Define the function

$$
w(x, p; \beta) = \frac{\partial v(x, p; \beta)}{\partial x} A(p)x + \left(\sum_{i=1}^{q} p_i \right)^2 \frac{\partial v(x, p; \beta)}{\partial p} \dot{p}.
\tag{5.15}
$$

We have that $w(x, p; \beta)$ is a parametrized form of degree $2m$ in x and degree $s+1$ in p. Moreover, it clearly follows that

$$\dot{v}(x, p; \beta) = w(x, p; \beta) \quad \forall p \in Y_q.
\tag{5.16}$$

If we choose $v(x, p; \beta)$ as candidate Lyapunov function, condition (5.8) in Theorem 5.1 can be rewritten as

$$
\begin{cases}
0 < v(x, \mathrm{sq}(p); \beta) \\
0 > w(x, \mathrm{sq}(p); \beta)
\end{cases}
\quad \forall x \in \mathbb{R}_0^n, \, \forall p \in \mathbb{R}_0^q, \, \forall \dot{p} \in \mathcal{D}.
\tag{5.17}
$$

Let us consider now $w(x, \mathrm{sq}(p); \beta)$. Let $R_1(\beta)$ and $R_2(\beta, \dot{p})$ be suitable symmetric matrices such that

$$
\begin{aligned}
\Psi\left(R_1(\beta), p^{\{s+1\}}, x^{\{m\}}\right) &= \frac{\partial v(x, \mathrm{sq}(p); \beta)}{\partial x} A(\mathrm{sq}(p))x \\
\Psi\left(R_2(\beta, \dot{p}), p^{\{s+1\}}, x^{\{m\}}\right) &= \left. \frac{\partial v(x, \sigma; \beta)}{\partial \sigma} \right|_{\sigma = \mathrm{sq}(p)} \left(\sum_{i=1}^{q} p_i^2 \right)^2 \dot{p}.
\end{aligned}
\tag{5.18}
$$

We hence have that $R_1(\beta) + R_2(\beta, \dot{p})$ is an SMR matrix of $w(x, \mathrm{sq}(p); \beta)$ with respect to the vector $p^{\{s+1\}} \otimes x^{\{m\}}$, i.e.

$$w(x, \mathrm{sq}(p); \beta) = \Psi\left(R_1(\beta) + R_2(\beta, \dot{p}), p^{\{s+1\}}, x^{\{m\}}\right). \tag{5.19}$$

Let us also observe that $R_1(\beta)$ depends linearly on β, while $R_2(\beta, \dot{p})$ is bilinear in β and $\dot{p}$, i.e. it is linear in β for fixed $\dot{p}$ and *vice versa*.

At this point we need to notice that the chosen SMR matrix $R_1(\beta) + R_2(\beta, \dot{p})$ is generally not unique. Indeed, one can also write

$$w(x, \mathrm{sq}(p); \beta) = \Psi\left(R_1(\beta) + R_2(\beta, \dot{p}) + L, p^{\{s+1\}}, x^{\{m\}}\right) \tag{5.20}$$

where L is any matrix belonging to the set $\mathscr{L}_{q,s+1,n,m}$, where

$$\mathscr{L}_{q,s,n,m} = \left\{ L \in \mathbb{S}^{\sigma(q,s)\sigma(n,m)} : \Psi\left(L, p^{\{s\}}, x^{\{m\}}\right) = 0 \ \ \forall x \in \mathbb{R}^n, \ \forall p \in \mathbb{R}^q \right\}. \tag{5.21}$$

The next result characterizes $\mathscr{L}_{q,s,n,m}$.

Theorem 5.4. *The set $\mathscr{L}_{q,s,n,m}$ is a linear space of dimension*

$$\omega(q,s,n,m) = \frac{1}{2}\sigma(q,s)\sigma(n,m)\left(\sigma(q,s)\sigma(n,m) + 1\right) - \sigma(q,2s)\sigma(n,2m). \tag{5.22}$$

Proof. Analogous to the proof of Theorem 1.2. $\square$

Table 5.1 reports $\omega(q,s,n,m)$ for some values of q,s,n,m. We are now ready to give a condition for establishing whether the robust stability property in Definition 5.2 holds.

Table 5.2 $\omega(q,s,n,m)$ for some values of q,s,n,m: (a) $q = 2$ and $n = 2$; (b) $q = 2$ and $n = 3$

	(a)					(b)			
	$s=1$	$s=2$	$s=3$	$s=4$		$s=1$	$s=2$	$s=3$	$s=4$
$m=1$	6	15	28	45	$m=1$	15	36	66	105
$m=2$	20	43	75	116	$m=2$	96	195	330	501
$m=3$	43	87	147	223	$m=3$	325	624	1023	1522
$m=4$	75	147	244	366	$m=4$	810	1515	2445	3600

Theorem 5.5. *The system (5.1)–(5.4) is robustly stable if there exist integers $s \geq 0$ and $m \geq 1$, $\beta \in \mathbb{R}^{\tau(q,s,n,m)}$ and $\alpha^{(1)}, \ldots, \alpha^{(h)} \in \mathbb{R}^{\omega(q,s+1,n,m)}$ such that*

$$\begin{cases} 0 < S(\beta) \\ 0 > R_1(\beta) + R_2(\beta, d^{(i)}) + L(\alpha^{(i)}), \quad i = 1, \ldots, h \end{cases} \tag{5.23}$$

where $S(\beta)$ is a linear parametrization of $\mathscr{S}_{q,s,n,m}$ in (5.11), $L(\cdot)$ is a linear parametrization of $\mathscr{L}_{q,s+1,n,m}$ in (5.21), and $R_1(\beta)$, $R_2(\beta, \cdot)$ satisfy (5.18).

Proof. Let $\beta, \alpha^{(1)}, \ldots, \alpha^{(h)}$ be such that (5.23) holds. Let us define $v(x, p; \beta)$ as in (5.13). From the first inequality of (5.23) and Theorem 5.3 one has that $v(x, p; \beta) > 0$ for all $x \in \mathbb{R}_0^n$ for all $p \in Y_q$. Then, from (5.15), (5.18) and the second inequality in (5.23) one has that

$$w(x, \mathrm{sq}(p); \beta) < 0 \quad \forall x \in \mathbb{R}_0^n, \ \forall p \in \mathbb{R}_0^q, \ \forall \dot{p} \in \left\{ d^{(1)}, \ldots, d^{(h)} \right\}.$$

Since $w(x, \mathrm{sq}(p); \beta)$ is affine in $\dot{p}$, and being $\mathscr{D}$ the convex hull of $d^{(1)}, \ldots, d^{(h)}$, it follows that

$$w(x, \mathrm{sq}(p); \beta) < 0 \quad \forall x \in \mathbb{R}_0^n, \ \forall p \in \mathbb{R}_0^q, \ \forall \dot{p} \in \mathscr{D}.$$

Hence, from Theorem 5.1 one can conclude that $v(x, p; \beta)$ is an HPD-HLF for the system (5.1)–(5.4), which is therefore robustly stable.   □

The total number of scalar variables involved in the LMI feasibility test (5.23) is equal to

$$\tau(q, s, n, m) + h\omega(q, s + 1, n, m).$$

Table 5.3 shows this quantity for some system dimensions.

Table 5.3 Total number of scalar parameters involved in (5.23) for some values of q, n, m, s: (a) $q = 2$ and $n = 2$; (b) $q = 2$ and $n = 3$

	(a)					(b)			
	$s=1$	$s=2$	$s=3$	$s=4$		$s=1$	$s=2$	$s=3$	$s=4$
$m=1$	19	45	83	133	$m=1$	45	105	192	306
$m=2$	56	121	213	332	$m=2$	255	531	915	1407
$m=3$	115	238	409	628	$m=3$	832	1657	2782	4207
$m=4$	196	396	671	1021	$m=4$	2040	3975	6585	9870

Example 5.1. Let us consider the system (5.1)–(5.4) with

$$A_1 = \begin{pmatrix} -1 & 1 \\ -2 & -1 \end{pmatrix}, \quad A_2 = \begin{pmatrix} 0 & 1 \\ -2 & -1 \end{pmatrix}, \quad d^{(1)} = \begin{pmatrix} 1 \\ -1 \end{pmatrix}, \quad d^{(2)} = \begin{pmatrix} -1 \\ 1 \end{pmatrix}.$$

Notice that (5.4) holds. Let us choose $m = 2$ and $s = 1$. The matrices involved in the robust stability condition (5.23) are

$$S(\beta) = \begin{pmatrix} \beta_1 & \beta_2 & \beta_3 & 0 & -\beta_{13} & -\beta_{14}-\beta_{15} \\ \star & \beta_4 & \beta_5 & \beta_{13} & \beta_{14} & -\beta_{16} \\ \star & \star & \beta_6 & \beta_{15} & 0 & \beta_{16} \\ \star & \star & \star & \beta_7 & \beta_8 & \beta_9 \\ \star & \star & \star & \star & \beta_{10} & \beta_{11} \\ \star & \star & \star & \star & \star & \beta_{12} \end{pmatrix}$$

$$L(\alpha) = \begin{pmatrix} 0 & 0 & -\alpha_1 & 0 & -\alpha_2 & -\alpha_3-\alpha_6 & -\alpha_{10} & -\alpha_4-\alpha_{11} & -l_1 \\ \star & 2\alpha_1 & 0 & \alpha_2 & \alpha_3 & -\alpha_7 & \alpha_4 & \alpha_5 & -\alpha_9-\alpha_{14} \\ \star & \star & 0 & \alpha_6 & \alpha_7 & 0 & \alpha_8 & \alpha_9 & -\alpha_{17} \\ \star & \star & \star & 2\alpha_{10} & \alpha_{11} & \alpha_{12} & 0 & -\alpha_{15} & -\alpha_{16}-\alpha_{18} \\ \star & \star & \star & \star & 2\alpha_{13} & \alpha_{14} & \alpha_{15} & \alpha_{16} & -\alpha_{19} \\ \star & \star & \star & \star & \star & 2\alpha_{17} & \alpha_{18} & \alpha_{19} & 0 \\ \star & \star & \star & \star & \star & \star & 0 & 0 & -\alpha_{20} \\ \star & \star & \star & \star & \star & \star & \star & 2\alpha_{20} & 0 \\ \star & \star & \star & \star & \star & \star & \star & \star & 0 \end{pmatrix}$$

$$l_1 = \alpha_5 + \alpha_8 + \alpha_{12} + \alpha_{13}$$

$$R_1(\beta) = \begin{pmatrix} r_1 & r_2 & r_3 & 0 & 0 & 0 & 0 & 0 & 0 \\ \star & 0 & r_4 & 0 & 0 & 0 & 0 & 0 & 0 \\ \star & \star & r_5 & 0 & 0 & 0 & 0 & 0 & 0 \\ \star & \star & \star & r_6 & r_7 & r_8 & 0 & 0 & 0 \\ \star & \star & \star & \star & 0 & r_9 & 0 & 0 & 0 \\ \star & \star & \star & \star & \star & r_{10} & 0 & 0 & 0 \\ \star & \star & \star & \star & \star & \star & r_{11} & r_{12} & r_{13} \\ \star & \star & \star & \star & \star & \star & \star & 0 & r_{14} \\ \star & \star & \star & \star & \star & \star & \star & \star & r_{15} \end{pmatrix}$$

$$R_2(\beta, d^{(i)}) = (-1)^{i-1} \begin{pmatrix} 1 & 0 & 0 \\ \star & 2 & 0 \\ \star & \star & 1 \end{pmatrix} \otimes \begin{pmatrix} r_{16} & r_{17} & r_{18} \\ \star & 0 & r_{19} \\ \star & \star & r_{20} \end{pmatrix}$$

$$\begin{aligned} r_1 &= -4\beta_1 - 2\beta_2 & r_2 &= 2\beta_1 - 2\beta_2 - 2\beta_3 - 2\beta_4 \\ r_3 &= 1.5\beta_2 - 2\beta_3 - 2\beta_4 - 3\beta_5 & r_4 &= \beta_3 + \beta_4 - 2\beta_5 - 4\beta_6 \\ r_5 &= \beta_5 - 4\beta_6 & r_6 &= -2\beta_2 - 4\beta_7 - 2\beta_8 \end{aligned}$$

$$r_7 = 2\beta_1 - 0.5\beta_2 - 2\beta_3 - 2\beta_4 + 2\beta_7 - 2\beta_8 - 2\beta_9 - 2\beta_{10}$$

$$r_8 = 1.5\beta_2 - \beta_3 - \beta_4 - 3\beta_5 + 1.5\beta_8 - 2\beta_9 - 2\beta_{10} - 3\beta_{11}$$

$$r_9 = \beta_3 + \beta_4 - 1.5\beta_5 - 4\beta_6 + \beta_9 + \beta_{10} - 2\beta_{11} - 4\beta_{12}$$

$$r_{10} = \beta_5 - 4\beta_6 + \beta_{11} - 4\beta_{12}$$

$$r_{11} = -2\beta_8 \qquad\qquad r_{12} = 2\beta_7 - 0.5\beta_8 - 2\beta_9 - 2\beta_{10}$$
$$r_{13} = 1.5\beta_8 - \beta_9 - \beta_{10} - 3\beta_{11} \qquad r_{14} = \beta_9 + \beta_{10} - 1.5\beta_{11} - 4\beta_{12}$$
$$r_{15} = \beta_{11} - 4\beta_{12} \qquad\qquad r_{16} = \beta_1 - \beta_7$$
$$r_{17} = 0.5(\beta_2 - \beta_8) \qquad\qquad r_{18} = 0.5(\beta_3 - \beta_9 + \beta_4 - \beta_{10})$$
$$r_{19} = 0.5(\beta_5 - \beta_{11}) \qquad\qquad r_{20} = \beta_6 - \beta_{12}.$$

The LMIs in (5.23) are feasible, and the found solution for β provides the HPD-HLF

$$v(x,p) = p_1(1.0000x_1^4 + 0.2936x_1^3x_2 + 1.0116x_1^2x_2^2 - 0.2816x_1x_2^3 + 0.5812x_2^4)$$
$$+ p_2(3.3074x_1^4 + 4.2908x_1^3x_2 + 3.5611x_1^2x_2^2 + 0.6306x_1x_2^3 + 0.6776x_2^4).$$

The following theorem states that the conservativeness of the condition of Theorem 5.5 does not increase by suitably increasing m and s.

Theorem 5.6. *If condition (5.23) of Theorem 5.5 holds for some integers $m \geq 1$ and $s \geq 0$, then it also holds with m and s replaced by km and $ks + l$ respectively, for all integers $k \geq 1$ and $l \geq 0$.*

Proof. First, we prove the theorem for $l = 0$. Let $\beta, \alpha^{(1)}, \ldots, \alpha^{(h)}$ be such that (5.23) holds for given m and s, and let $v(x, p; \beta)$ be the corresponding HPD-HLF, defined according to (5.13). We want to show that there exist $\tilde{\beta}, \tilde{\alpha}^{(1)}, \ldots, \tilde{\alpha}^{(h)}$ satisfying (5.23) with m and s replaced by km and ks, respectively. Let us introduce the new candidate Lyapunov function

$$\tilde{v}(x, p) = v(x, p; \beta)^k$$

which is a form of degree $2km$ in x and ks in p. One has that

$$\tilde{v}(x, \mathrm{sq}(p)) = \Psi\left(S(\beta), p^{\{s\}}, x^{\{m\}}\right)^k$$
$$= \Psi\left(X_1, p^{\{ks\}}, x^{\{km\}}\right)$$

where

$$X_1 = T_1' S(\beta)^{[k]} T_1$$

and T_1 is the matrix satisfying the relation

$$\left(p^{\{s\}} \otimes x^{\{m\}}\right)^{[k]} = T_1\left(p^{\{ks\}} \otimes x^{\{km\}}\right)$$

for all x, p. Let us observe that $X_1 \in \mathscr{S}_{q,s,n,m}$, as $\tilde{v}(x, \mathrm{sq}(p))$ does not contain monomials in p with odd powers. Moreover, $X_1 > 0$ because $S(\beta) > 0$ and T_1 is full column rank. Hence, there exists $\tilde{\beta}$ such that

$$\tilde{v}(x, \mathrm{sq}(p)) = \Psi\left(S(\tilde{\beta}), p^{\{ks\}}, x^{\{km\}}\right)$$

with $S(\tilde{\beta}) = X_1$, and therefore the first inequality in (5.23) holds.

Let us consider now the time derivative in each vertex $d^{(i)}$ of the polytope $\mathscr{D}$. By following the same reasoning as in (5.14)-(5.19), one gets

$$\dot{\bar{v}}(x, \mathrm{sq}(p)) = \Psi\left(R_1(\tilde{\beta}) + R_2(\tilde{\beta}, d^{(i)}), p^{\{ks+1\}}, x^{\{km\}}\right). \tag{5.24}$$

On the other hand, by using properties of Kronecker's products, one has that

$$\begin{aligned}
\dot{\bar{v}}(x, \mathrm{sq}(p)) &= k\Psi\left(S(\beta), p^{\{s\}}, x^{\{m\}}\right)^{k-1} \Psi\left(M^{(i)}, p^{\{s+1\}}, x^{\{m\}}\right) \\
&= k\left(\left(p^{\{s\}} \otimes x^{\{m\}}\right)' S(\beta)\left(p^{\{s\}} \otimes x^{\{m\}}\right)\right)^{k-1} \Psi\left(M^{(i)}, p^{\{s+1\}}, x^{\{m\}}\right) \\
&= k\left(p^{\{s\}} \otimes x^{\{m\}}\right)^{[k-1]'} S(\beta)^{[k-1]}\left(p^{\{s\}} \otimes x^{\{m\}}\right)^{[k-1]} \Psi\left(M^{(i)}, p^{\{s+1\}}, x^{\{m\}}\right) \\
&= k\left(\left(p^{\{s\}} \otimes x^{\{m\}}\right)^{[k-1]} \otimes \left(p^{\{s+1\}} \otimes x^{\{m\}}\right)\right)'\left(S(\beta)^{[k-1]} \otimes M^{(i)}\right) \\
&\qquad \left(\left(p^{\{s\}} \otimes x^{\{m\}}\right)^{[k-1]} \otimes \left(p^{\{s+1\}} \otimes x^{\{m\}}\right)\right) \\
&= \Psi\left(Y_1^{(i)}, p^{\{ks+1\}}, x^{\{km\}}\right)
\end{aligned}$$

$$\tag{5.25}$$

where

$$\begin{aligned}
M^{(i)} &= R_1(\beta) + R_2(\beta, d^{(i)}) + L(\alpha^{(i)}) \\
Y_1^{(i)} &= kT_2'\left(S(\beta)^{[k-1]} \otimes M^{(i)}\right) T_2
\end{aligned}$$

and T_2 is the matrix satisfying the relation

$$\left(p^{\{s\}} \otimes x^{\{m\}}\right)^{[k-1]} \otimes \left(p^{\{s+1\}} \otimes x^{\{m\}}\right) = T_2\left(p^{\{ks+1\}} \otimes x^{\{km\}}\right)$$

for all x, p. Hence, we have $Y_1^{(i)} < 0$ because $S(\beta) > 0$, $M^{(i)} < 0$ and T_2 is full column rank. Therefore, by (5.24) and (5.25) there exists $\tilde{\alpha}^{(i)}$ satisfying

$$R_1(\tilde{\beta}) + R_2(\tilde{\beta}, d^{(i)}) + L(\tilde{\alpha}^{(i)}) = Y_1^{(i)}, \quad i = 1, \dots, h$$

and the second inequality in (5.23) holds for every vertex $d^{(i)}$.

Now, we prove that the theorem holds also for $l > 0$. This can be done by showing that, if the condition of Theorem 5.5 is satisfied for $\tilde{m}$ and $\tilde{s}$, then it is also satisfied for $\tilde{m}$ and $\tilde{s}+1$. Let $\beta, \alpha^{(1)}, \dots, \alpha^{(h)}$ be the parameter vectors satisfying (5.23) for given $\tilde{m}$ and $\tilde{s}$, and be $v(x, p; \beta)$ the corresponding HPD-HLF. Let us define the new Lyapunov function

$$\tilde{v}(x, p) = v(x, p; \beta)\left(\sum_{i=1}^{q} p_i\right)$$

which is clearly a form of degree $2\tilde{m}$ in x and $\tilde{s}+1$ in p. It follows that

$$\tilde{v}(x, \text{sq}(p)) = \Psi\left(S(\beta), p^{\{\tilde{s}\}}, x^{\{\tilde{m}\}}\right)\left(\sum_{i=1}^{q} p_i^2\right)$$

$$= \Psi\left(X_2, p^{\{\tilde{s}+1\}}, x^{\{\tilde{m}\}}\right)$$

where $X_2 = U_1'\left(S(\beta) \otimes I_q\right)U_1$ and U_r is the matrix satisfying the relationship

$$p^{\{\tilde{s}+r-1\}} \otimes x^{\{\tilde{m}\}} \otimes p = U_r\left(p^{\{\tilde{s}+r\}} \otimes x^{\{\tilde{m}\}}\right)$$

for all x, p. Let us observe that $X_2 \in \mathscr{S}_{q, \tilde{s}+1, n, m}$, since $\tilde{v}(x, \text{sq}(p))$ does not contain monomials in p with odd powers, and $X_2 > 0$ because $S(\beta) > 0$ and U_1 is full column rank. Hence, there exists $\tilde{\beta}$ satisfying the first inequality of (5.23).

Finally, let us consider the time derivative in the vertex $d^{(i)}$. By using the second condition in (5.4), one has

$$\dot{\tilde{v}}(x, \text{sq}(p)) = \Psi\left(M^{(i)}, p^{\{\tilde{s}+1\}}, x^{\{\tilde{m}\}}\right)p'p + \tilde{v}(x, \text{sq}(p))\sum_{j=1}^{q} d_j^{(i)}$$

$$= \Psi\left(M^{(i)}, p^{\{\tilde{s}+1\}}, x^{\{\tilde{m}\}}\right)p'p$$

$$= \Psi\left(Y_2^{(i)}, p^{\{\tilde{s}+2\}}, x^{\{\tilde{m}\}}\right)$$

where $Y_2^{(i)} = U_2'(M^{(i)} \otimes I_q)U_2$. We have that $Y_2^{(i)} < 0$, because $M^{(i)} < 0$ and U_2 is full column rank. Therefore, there exists $\tilde{\alpha}^{(i)}$ satisfying the second inequality of (5.23) in the vertex $d^{(i)}$. This concludes the proof. $\square$

Several remarks can be made on the results provided by Theorems 5.5 and 5.6.

The condition of Theorem 5.5 is based on the parametrizations (5.10)–(5.11) and (5.20)–(5.21), and exploit the representation of the Lyapunov function and its time derivative via their SMR matrices. Since all possible SMR matrices of these two functions are considered by the introduced parametrization, the only source of conservatism in the resulting condition (5.23) originates from the gap between positive forms and SOS forms.

Theorem 5.6 guarantees that the conservatism of condition (5.23) in Theorem 5.5 does not increase, when one increases s for a fixed m. However, one can get more conservative results, if m is increased for a fixed $s \geq 1$ (see Example HPD-HLF-2 in Section 5.4).

Finally, it can be observed that conservatism is generally reduced by suitably increasing both m and s. This is guaranteed by Theorem 5.6, if both m and s are increased by the same factor (see also Example HPD-HLF-1 in Section 5.4). This confirms that it is useful to increase the degree of HPD-HLFs in both the state variables and the parameters when considering bounded-rate time-varying parametric uncertainty. Notice that this is not the case if the time derivative of the parameter $p(t)$ is unbounded, because in that case it is necessary to employ a common

Lyapunov function (i.e., parameter-independent, which corresponds to the case $s = 0$), as done in Chapter 3.

5.3   Robust Stability Margin

Hereafter we consider the problem of computing the robust stability margin defined as the maximum scaling factor of the polytope $\mathscr{D}$ for which robust stability is still guaranteed. Indeed, let us introduce

$$\mathscr{D}(\gamma) = \text{co}\left\{\gamma d^{(1)}, \dots, \gamma d^{(h)}\right\}. \tag{5.26}$$

We have the following definition.

Definition 5.4 (Maximum Variation Rate). Let us define

$$\gamma^{VR} = \sup\ \left\{\gamma \in \mathbb{R} : \text{(5.1) is robustly stable for all } p(t) \in \varUpsilon_q \text{ and } \dot{p}(t) \in \mathscr{D}(\gamma)\right\}. \tag{5.27}$$

Then, γ^{VR} is called *maximum variation rate* for the system (5.1).

The aim is to investigate γ^{VR} via HPD-HLFs. In particular, we define the maximum variation rate guaranteed by the class of HPD-HLFs of degree $2m$ in the state and degree s in the parameters as follows.

Definition 5.5 ($(2m, s)$-HPD-HLF Maximum Variation Rate). Let us define

$$\gamma_{2m,s}^{VR} = \sup\ \left\{\gamma \in \mathbb{R} :\ \ \exists v(x, p) \text{ HPD-HLF of degree } 2m \text{ in } x \text{ and } s \text{ in } p\ , \atop \text{for system (5.1), with } p(t) \in \varUpsilon_q,\ \dot{p}(t) \in \mathscr{D}(\gamma)\right\}. \tag{5.28}$$

Then, $\gamma_{2m,s}^{VR}$ is called $(2m, s)$-*HPD-HLF maximum variation rate* for the system (5.1).

Clearly, $\gamma_{2m,s}^{VR}$ is a lower bound of the sought maximum variation rate for all possible m, s, indeed

$$\gamma_{2m,s}^{VR} \leq \gamma^{VR} \quad \forall m \geq 1,\ \forall s \geq 0. \tag{5.29}$$

The following result provides a strategy for computing a lower bound of $\gamma_{2m,s}^{VR}$.

Theorem 5.7. *Let $s \geq 0$ and $m \geq 1$ be integers, $S(\beta)$ a linear parametrization of $\mathscr{S}_{q,s,n,m}$ in (5.11), and $L(\cdot)$ a linear parametrization of $\mathscr{L}_{q,s+1,n,m}$ in (5.21). Let us define*

$$\hat{\gamma}_{2m,s}^{VR} = \frac{1}{z^*} \tag{5.30}$$

where z^* is the solution of the GEVP

$$z^* = \inf_{z \in \mathbb{R},\ \beta \in \mathbb{R}^{\tau(q,s,n,m)},\ \alpha^{(0)},\dots,\alpha^{(h)} \in \mathbb{R}^{\omega(q,s+1,n,m)}} z$$

$$\text{s.t.} \begin{cases} 0 < S(\beta) \\ 0 < -R_1(\beta) - L(\alpha^{(0)}) \\ 0 < -z\left(R_1(\beta) + L(\alpha^{(0)})\right) - R_2(\beta, d^{(i)}) - L(\alpha^{(i)}), \\ \quad i = 1,\dots,h. \end{cases} \tag{5.31}$$

Then, $\hat{\gamma}_{2m,s}^{VR} \leq \gamma_{2m,s}^{VR}$.

Proof. Similarly to the proof of Theorem 5.5, the first constraint in (5.31) provides $v(x, p; \beta) > 0$ for all $x \in \mathbb{R}_0^n$ for all $p \in Y_q$. Then, one has

$$\dot{v}(x, \mathrm{sq}(p); \beta)|_{\dot{p} = z^{-1} d^{(i)}} = \Psi\left(R_1(\beta) + z^{-1} R_2(\beta, d^{(i)}), p^{\{s+1\}}, x^{\{m\}}\right)$$
$$= z^{-1} \Psi\left(z R_1(\beta) + R_2(\beta, d^{(i)}), p^{\{s+1\}}, x^{\{m\}}\right).$$

Since $L(\alpha^{(0)}), \dots, L(\alpha^{(h)}) \in \mathscr{L}_{q,s+1,n,m}$, it follows that

$$\dot{v}(x, \mathrm{sq}(p); \beta)|_{\dot{p} = z^{-1} d^{(i)}}$$
$$= z^{-1} \Psi\left(z\left(R_1(\beta) + L(\alpha^{(0)})\right) - R_2(\beta, d^{(i)}) - L(\alpha^{(i)}), p^{\{s+1\}}, x^{\{m\}}\right)$$

and hence the third constraint in (5.31) ensures that

$$\dot{v}(x, \mathrm{sq}(p); \beta)|_{\dot{p} = z^{-1} d^{(i)}} < 0 \quad \forall x \in \mathbb{R}_0^n, \forall p \in \mathbb{R}_0^q.$$

From Theorem 1.17 one can hence conclude that

$$\dot{v}(x, p; \beta) < 0 \quad \forall x \in \mathbb{R}_0^n, \forall p \in Y_q, \forall \dot{p} \in z^{-1} D$$

and therefore $\hat{\gamma}_{2m,s}^{VR} \leq \gamma_{2m,s}^{VR}$. $\qquad\square$

Theorem 5.7 states that a lower bound to $\gamma_{2m,s}^{VR}$ can be computed through the GEVP (5.31), which is a quasi-convex optimization problem. This is therefore a lower bound also to the actual maximum variation rate γ^{VR} due to (5.29). It is worth observing that the second inequality in (5.31), which is required in order to guarantee that the optimization is a GEVP, does not reduce the set of feasible α since $0_q \in \mathscr{D}$.

5.4 Examples

In this section we present some examples concerning robustness of uncertain systems with time-varying structured parametric uncertainty and bounded variation rate.

5.4.1 Example HPD-HLF-1

Let us consider the following system, which is quite popular in the literature ([148, 146, 1, 34, 93]),

$$\dot{x}(t) = \begin{pmatrix} 0 & 1 \\ -2 - r(t) & -1 \end{pmatrix} x(t), \quad 0 \le r(t) \le k, \ |\dot{r}(t)| \le \gamma. \qquad (5.32)$$

We want to compute the maximum value of γ such that (5.32) is asymptotically stable for all admissible functions $r(t)$, for a given value of k.

Let us start by rewriting the system (5.32) as in (5.1)-(5.4), for $\gamma = 1$. This can be done by selecting

$$A_1 = \begin{pmatrix} 0 & 1 \\ -2 & -1 \end{pmatrix}, \quad A_2 = \begin{pmatrix} 0 & 1 \\ -2-k & -1 \end{pmatrix}$$

$$d^{(1)} = \frac{1}{k} \begin{pmatrix} 1 \\ -1 \end{pmatrix}, \quad d^{(2)} = \frac{1}{k} \begin{pmatrix} -1 \\ 1 \end{pmatrix}.$$

The problem boils down to the computation of the maximum variation rate γ^{VR} in (5.27). Lower bounds of γ^{VR} can be computed by using HPD-HLFs and Theorem 5.7. Figures 5.1a–b and 5.2a show the results obtained with $m = 1, 2, 3$ and $s = 0, 1, 2, 3$ for some values of k.

We also observe that robust stability against unbounded $\dot{p}(t)$ can be guaranteed for $k = 6.8649$ by using an HPD-HLF with $m = 10$ and $s = 0$, i.e. an HPLF (see Example HPLF-1 in Chapter 3).

As one can see, the conservativeness of the condition of Theorem 5.5 does not increase by increasing s for a fixed m, but may increase by increasing m for a fixed s. This is clearly shown in Figure 5.2b which reports the lower bounds for $s = 1$ and $m = 1, 2, 3$.

5.4.2 Example HPD-HLF-2

In order to show another feature of HPD-HLFs, let us consider system (5.32) for $k = 110$ and $\dot{r}(t) \equiv 0$, which is obviously asymptotically stable. Its robust stability can be ensured by using Theorem 5.5 with $m = 1$ and $s = 1$, in particular an HPD-HLF is $v(x, p) = x'(p_1 V_1 + p_2 V_2)x$ with

$$V_1 = \begin{pmatrix} 1 & 0.1429 \\ \star & 0.4286 \end{pmatrix}, \quad V_2 = \begin{pmatrix} 47.6085 & 0.0206 \\ \star & 0.4325 \end{pmatrix}.$$

However, robust stability cannot be established by any HPD-HLF with $m = 2$ and $s = 1$. Indeed, there does not even exist any linearly parameter-dependent quartic

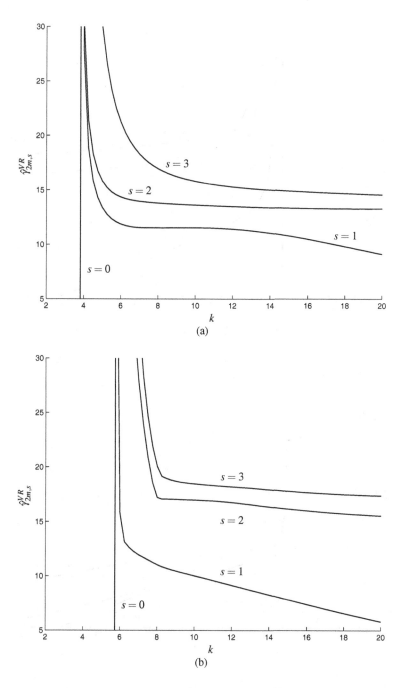

Fig. 5.1 Example HPD-HLF-1, lower bound $\hat{\gamma}_{2m,s}^{VR}$ versus parameter bound k for $s = 0, 1, 2, 3$: (a) $m = 1$; (b) $m = 2$

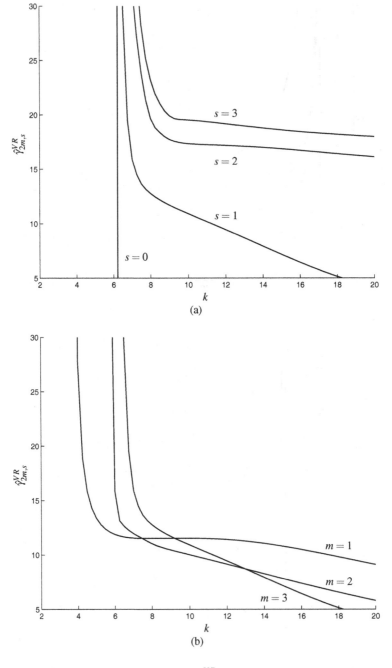

Fig. 5.2 Example HPD-HLF-1, lower bound $\hat{\gamma}_{2m,s}^{VR}$ versus parameter bound k: (a) $m = 3$ for $s = 0, 1, 2, 3$; (b) $s = 1$ for $m = 1, 2, 3$

Lyapunov function simultaneously verifying the stability of the matrix pencil $A(p)$ at the points

$$p^{(1)} = \begin{pmatrix} 1 \\ 0 \end{pmatrix}, \quad p^{(2)} = \frac{1}{2} \begin{pmatrix} 1 \\ 1 \end{pmatrix}, \quad p^{(3)} = \begin{pmatrix} 0 \\ 1 \end{pmatrix}.$$

In fact, let us write such a candidate Lyapunov function as

$$v(x,p) = x^{\{2\}'} \left(p_1 \tilde{V}_1 + p_2 \tilde{V}_2 \right) x^{\{2\}}$$

where $\tilde{V}_1, \tilde{V}_2 \in \mathbb{R}^{3\times 3}$ are symmetric matrices. Then, $v(x,p)$ should satisfy the following conditions:

$$\begin{cases} 0 < v(x,p) \\ 0 > \dot{v}(x,p) \end{cases} \quad \forall x \neq 0_n, \ \forall p \in \left\{ p^{(1)}, p^{(2)}, p^{(3)} \right\} \qquad (5.33)$$

where

$$\dot{v}(x,p) = x^{\{2\}'} R(p) x^{\{2\}}$$
$$R(p) = p_1^2 \mathrm{he}\left(\tilde{V}_1 A_1^{\#} \right) + p_1 p_2 \mathrm{he}\left(\tilde{V}_1 A_2^{\#} + \tilde{V}_2 A_1^{\#} \right) + p_2^2 \mathrm{he}\left(\tilde{V}_2 A_2^{\#} \right)$$

being $A_i^{\#} \in \mathbb{R}^{3\times 3}$ the extended matrix of A_i defined by (3.8) for $m = 2$. For a fixed p, $v(x,p)$ and $\dot{v}(x,p)$ are forms of degree 4 in two variables (x_1 and x_2). Such forms are positive if and only if they admit a positive definite SMR matrix according to Theorem 2.4. Therefore, (5.33) is satisfied if and only if there exist $\tilde{V}_1, \tilde{V}_2$ and $\alpha^{(1)}, \alpha^{(2)}, \alpha^{(3)} \in \mathbb{R}$ such that

$$\begin{cases} 0 < \tilde{V}_i \quad \forall i = 1,2 \\ 0 > \mathrm{he}\left(\tilde{V}_i A_i^{\#} \right) + L(\alpha^{(i)}) \quad \forall i = 1,2 \\ 0 > \mathrm{he}\left((\tilde{V}_1 + \tilde{V}_2)(A_1^{\#} + A_2^{\#}) \right) + L(\alpha^{(3)}) \end{cases} \qquad (5.34)$$

where $L(\cdot)$ is a linear parametrization of $\mathscr{L}_{2,4}$ in (1.15). Since it can be checked that the set of LMIs (5.34) is infeasible, one can conclude that the sought linearly parameter-dependent quartic Lyapunov function satisfying (5.33) does not exist.

5.4.3 Example HPD-HLF-3

Let us consider the following system, borrowed from [93, 14],

$$\dot{x}(t) = \left(\begin{pmatrix} 8 & -9 \\ 120 & -18 \end{pmatrix} + \begin{pmatrix} -108 & 9 \\ -120 & 17 \end{pmatrix} p(t) \right) x(t), \quad 0 \leq p(t) \leq 1, \ |\dot{p}(t)| \leq \gamma.$$

We want to find the maximum value of γ such that the system is asymptotically stable for all admissible functions $p(t)$.

The problem requires to compute the maximum variation rate γ^{VR} in (5.27). By using Theorem 5.7 for $m = 2$ and $s = 0$ we find

$$\hat{\gamma}^{VR} = +\infty = \gamma^{VR}.$$

On the other hand, for $m = 1$ and $s = 1$ we find $\hat{\gamma}^{VR} = 63.25$. This clearly shows that homogenous polynomial Lyapunov functions (even with no parameter dependence) can provide better results than affine parameter dependent quadratic Lyapunov functions.

5.4.4 Example HPD-HLF-4

Let us consider the system

$$\dot{x}(t) = A(p(t),k),x(t), \quad p(t) \in \Upsilon_q, \ \dot{p}(t) \in \mathscr{D}$$

where

$$A(p,k) = \sum_{i=1}^{3} p_i \bar{A}_i(k), \quad \bar{A}_i(k) = \bar{A}_0 + k\bar{A}_i$$

and the matrices $\bar{A}_0, \ldots, \bar{A}_3$ are chosen as

$$\bar{A}_0 = \begin{pmatrix} -2.4 & -0.6 & -1.7 & 3.1 \\ 0.7 & -2.1 & -2.6 & -3.6 \\ 0.5 & 2.4 & -5 & -1.6 \\ -0.6 & 2.9 & -2 & -0.6 \end{pmatrix}, \quad \bar{A}_1 = \begin{pmatrix} 1.1 & -0.6 & -0.3 & -0.1 \\ -0.8 & 0.2 & -1.1 & 2.8 \\ -1.9 & 0.8 & -1.1 & 2 \\ -2.4 & -3.1 & -3.7 & -0.1 \end{pmatrix}$$

$$\bar{A}_2 = \begin{pmatrix} 0.9 & 3.4 & 1.7 & 1.5 \\ -3.4 & -1.4 & 1.3 & 1.4 \\ 1.1 & 2 & -1.5 & -3.4 \\ -0.4 & 0.5 & 2.3 & 1.5 \end{pmatrix}, \quad \bar{A}_3 = \begin{pmatrix} -1 & -1.4 & -0.7 & -0.7 \\ 2.1 & 0.6 & -0.1 & -2.1 \\ 0.4 & -1.4 & 1.3 & 0.7 \\ 1.5 & 0.9 & 0.4 & -0.5 \end{pmatrix}.$$

The set $\mathscr{D}$ in given by (5.3) with

$$d^{(1)} = \begin{pmatrix} 1 \\ 0 \\ -1 \end{pmatrix}, \quad d^{(2)} = \begin{pmatrix} -1 \\ 0 \\ 1 \end{pmatrix}.$$

Notice that this corresponds to the constraints

$$|\dot{p}_1| \le 1, \quad \dot{p}_2 = 0, \quad |\dot{p}_3| \le 1, \quad \dot{p}_1 + \dot{p}_3 = 0$$

i.e., one time-invariant parameter and two time-varying parameters on the unit simplex. The aim is to compute the maximum variation rate γ^{VR} in (5.27) for different values of k.

The lower bound $\hat{\gamma}^{VR}_{2m,s}$ provided by HPD-HLFs with $m = s = 1$ (348 LMI parameters) are plotted against k in Figure 5.3. Figure 5.3 reports also the curve corresponding to HPD-HLFs with $m = 1$, $s = 2$ (1290 LMI parameters), which confirms the benefits of increasing the degree s of the form in the uncertain parameter p.

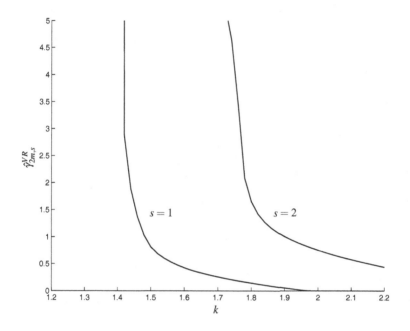

Fig. 5.3 Example HPD-HLF-4, lower bound $\hat{\gamma}^{VR}_{2m,s}$ versus k, for $m = 1$ and $s = 1, 2$

5.5 Notes and References

HPD-HLFs have been introduced in [37]. The main robust stability results in this chapter have been presented in [43]. The model (5.1)–(5.4) for systems affected by bounded-rate time-varying uncertainty is the same considered in [68], and it can be seen as an extension of the models adopted in previous works such as [65] and [94].

Robust stability conditions for polytopic systems affected by time-varying uncertainties with bounds on the variation rate, have been derived by several authors. In [65], affine parameter-dependent quadratic Lyapunov functions have been employed, while in [94, 11, 145, 68, 12] different relaxations based on quadratic Lyapunov functions with polynomial dependence on the parameters have been proposed.

Chapter 6
Distance Problems with Applications to Robust Control

This chapter presents further results for robustness analysis of uncertain systems based on forms. It is shown how the problem of computing the euclidean distance from a point to a surface described by a polynomial equation, can be solved via LMI feasibility tests. This problem has numerous applications in systems and control theory. In this respect, we consider the computation of the l_2 parametric stability margin of systems affected by time-invariant uncertainty.

6.1 Quadratic Distance Problems

Let us start by introducing the following definition.

Definition 6.1 (QDP). Let $f(x)$ be a polynomial of degree m, and let $Q \in \mathbb{R}^{n \times n}$ be a positive definite matrix. Then, the optimization problem

$$
\begin{aligned}
c_{min} = \inf_{x \in \mathbb{R}^n} \ & x'Qx \\
\text{s.t.} \ & f(x) = 0
\end{aligned}
\tag{6.1}
$$

is called a *QDP*.

From a geometric point of view, a QDP consists of the computation of the minimum weighted euclidean distance of the origin from a surface defined by the polynomial constraint $f(x) = 0$. Equivalently, it amounts to find the largest ellipsoid centered in the origin, with shape matrix Q, contained in the region bounded by $f(x) = 0$.

Without loss of generality, we can assume

$$
f(0_n) \neq 0.
\tag{6.2}
$$

Indeed, if the origin is feasible for the constraint $f(x) = 0$, then the solution of the QDP (6.1) is trivially $c_{min} = 0$ and one can solve the problem by simply checking

G. Chesi et al.: Homogeneous Polynomial Forms, LNCIS 390, pp. 155–175.
springerlink.com   © Springer-Verlag Berlin Heidelberg 2009

whether $f(0_n) = 0$ or not. Let us observe that, being $f(x)$ a continuous function, (6.2) implies that we can assume without loss of generality that $f(x)$ is locally positive definite, i.e.

$$\exists \delta > 0 : f(x) > 0 \quad \forall x \in \mathbb{R}^n, \|x\| < \delta. \tag{6.3}$$

Obviously, if $f(x)$ is locally negative definite, one can simply redefine $f(x)$ as $-f(x)$ without altering the solution of (6.1).

Notice that QDPs are in general nonconvex optimization problems. This means that they may admit different locally optimal solutions, as shown in the following example.

Example 6.1. Let us consider problem (6.1) with

$$f(x) = 1 + x_1 - x_2^2 + 2x_1^2 x_2$$
$$Q = I_2.$$

Figure 6.1 shows the feasible set corresponding to the constraint $f(x) = 0$, the globally optimal solution corresponding to the smallest circle, and a locally optimal solution corresponding to the largest circle.

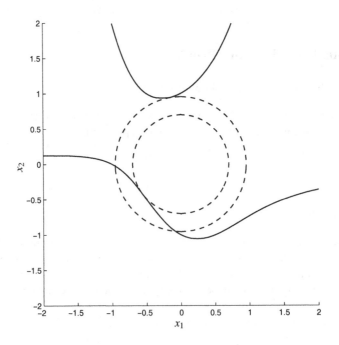

Fig. 6.1 Example 6.1: constraint $f(x) = 0$ (solid line), global minimum (inner dashed circle), and local minimum (outer dashed circle)

In the following, it will be shown how QDPs can be solved by exploiting the results of Section 1.7 on positive polynomials on ellipsoids. Let $\mathcal{B}(Q,c)$ be the ellipsoid defined in (1.60), and observe that, for sufficiently small c, assumption (6.3) ensures that $\mathcal{B}(Q,c)$ does not intersect the constraint set $\{x : f(x) = 0\}$. This suggests that the solution of a QDP can be computed via a one-parameter family of "cutting" tests. Specifically, the following lemma holds.

Lemma 6.1. *The solution c_{min} of problem (6.1) is given by*

$$c_{min} = \sup \ \{\hat{c} \in \mathbb{R} : f(x) > 0 \ \forall x \in \mathcal{B}(Q,c), \ \forall c \in (0,\hat{c}]\}. \tag{6.4}$$

Proof. Due to assumption (6.3), $f(x) > 0$ for all $x \in \mathcal{B}(Q,c)$, for sufficiently small c. Moreover, since the constraint set $f(x) = 0$ does not intersect ellipsoids $\mathcal{B}(Q,c)$ with $c < c_{min}$, it follows that $f(x)$ is strictly positive on these level surfaces. $\qquad\square$

Lemma 6.1 asserts that c_{min} can be found by solving a one-parameter family of positivity tests on polynomial $f(x)$, for x belonging to a given ellipsoid $\mathcal{B}(Q,c)$. Notice that the constrained positivity condition in (6.4) is exactly the same as in (1.59). By using Theorem 1.15 one can get rid of the set $\mathcal{B}(Q,c)$ and obtain c_{min} by performing positivity tests on unconstrained forms $w(x;c)$, i.e.

$$c_{min} = \sup \ \{\hat{c} \in \mathbb{R} : w(x;c) > 0 \ \forall x \in \mathbb{R}_0^n, \ \forall c \in (0,\hat{c}]\} \tag{6.5}$$

where $w(x;c)$ takes on either one of the following expressions:

1. if $f(x)$ is a generic polynomial (not even),

$$w(x;c) = \sum_{i=0}^{m} \bar{f}_{2i}(x) \left(\frac{x'Qx}{c} \right)^{m-i} \tag{6.6}$$

where $\bar{f}_{2i} \in \Xi_{n,2i}$ satisfy

$$f(x)f(-x) = \sum_{i=0}^{m} \bar{f}_{2i}(x); \tag{6.7}$$

2. if $f(x)$ is an even polynomial,

$$w(x;c) = \sum_{i=0}^{m} f_{2i}(x) \left(\frac{x'Qx}{c} \right)^{m-i} \tag{6.8}$$

where $f_{2i} \in \Xi_{n,2i}$ satisfy

$$f(x) = \sum_{i=0}^{m} f_{2i}(x). \tag{6.9}$$

Then, by using Theorem 1.16 one can obtain a lower bound on c_{min} via LMI feasibility tests as described in the following result.

Theorem 6.1. *Let us define*

$$\hat{c}_{min} = \sup \ \{\hat{c}: \quad either \ condition \ 1 \ or \ condition \ 2 \ in \tag{6.10}$$
$$Theorem \ 1.16 \ holds, \ \forall c \in (0, \hat{c}]\}.$$

Then, $\hat{c}_{min} \leq c_{min}$.

Proof. First notice that (1.61) holds for all $c < c_{min}$, otherwise c_{min} is not the solution of (6.1). Hence, the treatment in Section 1.7 can be applied: in particular, conditions 1 or 2 in Theorem 1.16 guarantee that $f(x) > 0$ for all $x \in \mathscr{B}(Q, c)$, for a given c. Therefore, the result follows by Lemma 6.1.   □

Theorem 6.1 suggests that a lower bound to the minimum in (6.1) can be approximated within the desired precision by solving a one-parameter family of LMI feasibility tests, generated by performing a sweeping with respect to the parameter c. Indeed, let us make the following assumption

$$\exists \varepsilon > 0 : \ \forall \tilde{c} \in (c_{min}, c_{min} + \varepsilon), \ \exists x \in \mathscr{B}(Q, \tilde{c}) \ \text{such that} \ f(x) < 0$$

i.e., we assume that the constraint $f(x)$ takes negative values on ellipsoids $\mathscr{B}(Q, \tilde{c})$, with $\tilde{c}$ in a right neighborhood of c_{min}. Then, by performing a sweeping on c with step smaller than ε, the one-parameter family of LMI feasibility tests resulting from (6.10) provides a guaranteed lower bound to c_{min}, according to Theorem 6.1.

Example 6.2. Let us consider the QDP in Example 6.1. We have that $f(x)$ is a polynomial of degree $m = 3$, hence we use condition 1 in Theorem 1.16 and we find

$$\hat{c}_{min} = 0.4901.$$

Figure 6.2 shows the corresponding circle $x_1^2 + x_2^2 = \hat{c}_{min}$. Let us observe also that this curve is tangent to the constraint $f(x) = 0$, hence implying that the found lower bound is tight. Figure 6.3 shows the constraint $f(x)f(-x) = 0$ used to generate the parametrized form $w(x; c)$ according to (6.6)–(6.7).

6.2 Special Cases and Extensions

In this section, we consider a special class of QDPs and we extend the proposed techniques to maximum distance problems.

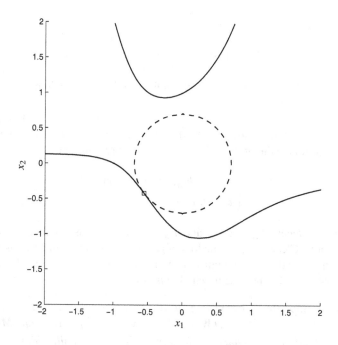

Fig. 6.2 Example 6.2: constraint $f(x) = 0$ (solid line) and set $\mathscr{B}(Q, \hat{c}_{min})$ (dashed line)

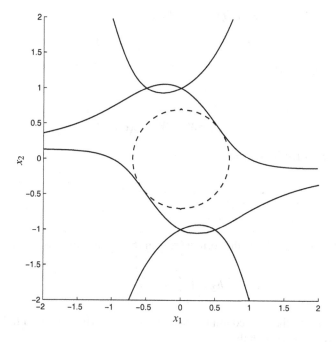

Fig. 6.3 Example 6.2: constraint $f(x)f(-x) = 0$ (solid line) and set $\mathscr{B}(Q, \hat{c}_{min})$ (dashed line)

6.2.1   The Two-form Case

A special case of the QDP (6.1) occurs when the constraint function $f(x)$ is given by the sum of two forms of different degrees. Indeed, in this case the lower bound $\hat{c}_{min}$ given by (6.10) can be computed through a single LMI problem.

Before proceeding, let us observe that the case in which $f(x)$ is itself a form does not need to be considered. In fact, since (6.3) holds, it directly follows that $f(x) > 0$ for all x, and hence the constraint in QDP (6.1) does not admit any feasible solution.

Let us reformulate the problem as in (6.1), by expressing $f(x)$ as

$$\begin{cases} f(x) = h_{2a}(x) + h_{2a+b}(x) \\ a \ge 0, \ b > 0 \end{cases} \tag{6.11}$$

where a, b are integers, $h_{2a} \in \Xi_{n,2a}$, and $h_{2a+b} \in \Xi_{n,2a+b}$. It is assumed that $h_{2a}(x)$ is positive definite. Clearly, there is no loss of generality in introducing this assumption since (6.3) holds and $h_{2a}(x)$ contains the monomials of lowest degree of $f(x)$.

Let us first consider the case in which b is odd.

Theorem 6.2. *Suppose that $f(x)$ is as in (6.11) with b odd, and let us denote the degree of $f(x)$ by $m = 2a + b$. Let W_1 and $W_2 + L(\alpha)$ be respectively an SMR matrix of $h_{2a}(x)^2 (x'Qx)^b$ and a complete SMR matrix of $h_{2a+b}(x)^2$, built with respect to the same power vector $x^{\{m\}}$, i.e.*

$$h_{2a}(x)^2 \left(x'Qx \right)^b = x^{\{m\}'} W_1 x^{\{m\}} \tag{6.12}$$

$$h_{2a+b}(x)^2 = x^{\{m\}'} (W_2 + L(\alpha)) x^{\{m\}}. \tag{6.13}$$

Let us define the EVP

$$z^* = \inf_{z \in \mathbb{R}, \ \alpha \in \mathbb{R}^{\omega(n,m)}} z \tag{6.14}$$

$$s.t. \ \ zW_1 - W_2 - L(\alpha) > 0.$$

Then, $\hat{c}_{min}$ in (6.10) is given by

$$\hat{c}_{min} = z^{* - \frac{1}{b}}. \tag{6.15}$$

Proof. We first observe that the form $w(x;c)$ in (6.6) boils down to

$$w(x;c) = h_{2a}(x)^2 \left(\frac{x'Qx}{c} \right)^b - h_{2a+b}(x)^2.$$

Hence, it follows that the condition $\lambda(w(\cdot;c)) > 0$ in Theorem 1.16 holds if and only if there exists α such that

$$\frac{W_1}{c^b} - W_2 - L(\alpha) > 0$$

and therefore one gets (6.14)–(6.15). $\qquad\square$

For the case in which b is even we have the following result.

Theorem 6.3. *Suppose that $f(x)$ is as in (6.11) with b even, and let us denote the degree of $f(x)$ by $2m = 2a + b$. Let W_1 and $W_2 + L(\alpha)$ be respectively an SMR matrix of $h_{2a}(x)(x'Qx)^{\frac{b}{2}}$ and a complete SMR matrix of $h_{2a+b}(x)$, built with respect to the same power vector $x^{\{m\}}$, i.e.*

$$h_{2a}(x)(x'Qx)^{\frac{b}{2}} = x^{\{m\}'} W_1 x^{\{m\}} \tag{6.16}$$

$$h_{2a+b}(x) = x^{\{m\}'}(W_2 + L(\alpha)) x^{\{m\}}. \tag{6.17}$$

Let us define the EVP

$$z^* = \inf_{z \in \mathbb{R}, \; \alpha \in \mathbb{R}^{\omega(n,m)}} z \tag{6.18}$$
$$\text{s.t. } zW_1 + W_2 + L(\alpha) > 0.$$

Then, $\hat{c}_{min}$ in (6.10) is given by

$$\hat{c}_{min} = z^{* -\frac{2}{b}}. \tag{6.19}$$

Proof. Similar to the proof of Theorem 6.2 by observing that the form $w(x;c)$ in (6.8) boils down to

$$w(x;c) = h_{2a}(x)\left(\frac{x'Qx}{c}\right)^{\frac{b}{2}} + h_{2a+b}(x) \tag{6.20}$$

and that the condition $\lambda(w(\cdot;c)) > 0$ in Theorem 1.16 holds if and only if there exists α such that
$$\frac{W_1}{c^{\frac{b}{2}}} + W_2 + L(\alpha) > 0.$$

Therefore, $\hat{c}_{min}$ is given (6.18)–(6.19). $\qquad\square$

Example 6.3. Let us consider problem (6.1) with

$$f(x) = 1 + x_1^3 - 2x_1^2 x_2 + 3x_1 x_2^2 - 4x_2^3$$
$$Q = \begin{pmatrix} 1 & 0.5 \\ \star & 2 \end{pmatrix}.$$

Observe that $f(x)$ is as in (6.11) for

$$a = 0, \quad b = 3$$
$$h_{2a}(x) = 1$$
$$h_{2a+b}(x) = x_1^3 - 2x_1^2 x_2 + 3x_1 x_2^2 - 4x_2^3.$$

We solve the EVP (6.14) and find from (6.15) the solution

$$\hat{c}_{min} = 0.4219.$$

Figure 6.4 shows the constraint $f(x) = 0$ and the ellipse $\mathscr{B}(Q, \hat{c}_{min})$. Observe that the found lower bound is tight.

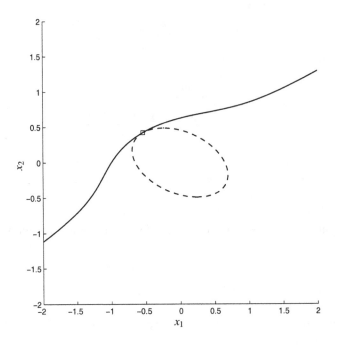

Fig. 6.4 Example 6.3: constraint $f(x) = 0$ (solid line) and set $\mathscr{B}(Q, \hat{c}_{min})$ (dashed line)

6.2.2 Maximum QDPs

All the treatment described in the previous sections for addressing QDPs, can be extended to QDPs where the target is computing the maximum rather than the minimum of the function $x'Qx$, i.e.

$$c_{max} = \sup_{x \in \mathbb{R}^n} x'Qx$$
$$\text{s.t. } f(x) = 0 \tag{6.21}$$

where Q is a positive definite matrix and $f(x)$ is a polynomial of degree m.

From a geometric point of view, problem (6.21) consists of computing the smallest ellipsoid centered in the origin, with shape matrix Q, that contains the surface defined by the polynomial constraint $f(x) = 0$.

Without loss of generality, we can assume that $f(x)$ satisfies the following definiteness property:

$$\exists \delta > 0 : f(x) > 0 \quad \forall x \in \mathbb{R}^n, \|x\| > \delta^{-1}. \tag{6.22}$$

In fact, if this assumption does not hold, then either one of the following situations occur:

1. for all $\delta > 0$ there exists $x \in \mathbb{R}^n$ such that $\|x\| > \delta^{-1}$ and $f(x) = 0$, which implies that the solution of (6.21) is $c_{max} = \infty$;
2. there exists $\delta > 0$ such that $f(x) < 0$ for all $x \in \mathbb{R}^n$ with $\|x\| > \delta^{-1}$, which implies that one can redefine $f(x)$ as $-f(x)$ in order to fulfill (6.22) without altering the solution of (6.21).

By using the same arguments as in Section 6.1, one obtains

$$c_{max} = \inf \{\hat{c} \in \mathbb{R} : f(x) > 0 \quad \forall x \in \mathcal{B}(Q,c), \forall c \in [\hat{c}, \infty)\}. \tag{6.23}$$

The following result provides the counterpart of Theorem 6.1 for the maximum QDP (6.21), and shows how an upper bound of c_{max} can be obtained via LMI feasibility tests.

Theorem 6.4. *Let us define*

$$\hat{c}_{max} = \inf \{\hat{c} : \textit{either condition 1 or condition 2 of}$$
$$\textit{Theorem 1.16 holds, } \forall c \in [\hat{c}, \infty)\}. \tag{6.24}$$

Then, $\hat{c}_{max} \geq c_{max}$.

Example 6.4. Let us consider problem (6.21) with

$$f(x) = -1 - x_1^2 + 2x_1x_2 + x_2^2 + x_1^4 + x_2^4$$
$$Q = \begin{pmatrix} 2 & -0.5 \\ \star & 1 \end{pmatrix}.$$

We have that $f(x)$ is an even polynomial of degree $m = 4$, hence we use condition 2 in Theorem 1.16 and we find

$$\hat{c}_{max} = 6.2320.$$

Figure 6.5 shows the corresponding ellipse $\mathscr{B}(Q, \hat{c}_{max})$ and the constraint $f(x) = 0$. It can be observed that the curve is tangent to the constraint $f(x) = 0$, hence implying that the found upper bound is tight.

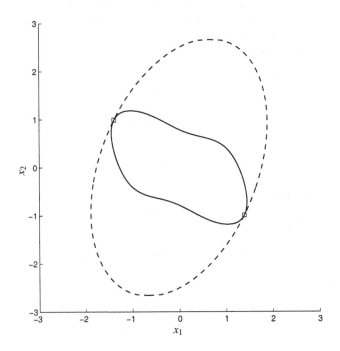

Fig. 6.5 Example 6.4: constraint $f(x) = 0$ (solid line) and $\mathscr{B}(Q, \hat{c}_{max})$ (dashed line)

Also for maximum QDPs it is possible to simplify the computation of the sought bound if $f(x)$ is given by the sum of two forms of different degrees. Indeed, let us suppose that

$$\begin{cases} f(x) = h_{2a-b}(x) + h_{2a}(x) \\ 2a - b \geq 0, \ b > 0 \end{cases} \tag{6.25}$$

where a, b are integers, $h_{2a-b} \in \Xi_{n,2a-b}$, and $h_{2a} \in \Xi_{n,2a}$. It is assumed that $h_{2a}(x)$ is positive definite. Again, there is no loss of generality in introducing this assumption since (6.22) holds and $h_{2a}(x)$ contains the monomials of highest degree of $f(x)$.

Depending on the parity of b, we have the following two results, whose proof are analogous to those of Theorems 6.2 and 6.3.

Theorem 6.5. *Suppose that $f(x)$ is as in (6.25) with b odd, and let us denote the degree of $f(x)$ by $m = 2a$. Let W_1 and $W_2 + L(\alpha)$ be respectively an SMR matrix of $h_{2a}(x)^2$ and a complete SMR matrix of $h_{2a-b}(x)^2 (x'Qx)^b$, built with respect to the same power vector $x^{\{m\}}$, i.e.*

$$h_{2a}(x)^2 = x^{\{m\}'} W_1 x^{\{m\}}$$

$$h_{2a-b}(x)^2 (x'Qx)^b = x^{\{m\}'} (W_2 + L(\alpha)) x^{\{m\}}.$$

Let us define the EVP

$$z^* = \inf_{z \in \mathbb{R}, \, \alpha \in \mathbb{R}^{\omega(n,m)}} z \tag{6.26}$$

$$\text{s.t. } zW_1 - W_2 - L(\alpha) > 0.$$

Then, $\hat{c}_{max}$ in (6.24) is given by

$$\hat{c}_{max} = z^{*\frac{1}{b}}. \tag{6.27}$$

Theorem 6.6. *Suppose that $f(x)$ is as in (6.25) with b even, and let us denote the degree of $f(x)$ by $2m = 2a$. Let W_1 and $W_2 + L(\alpha)$ be respectively an SMR matrix of $h_{2a}(x)$ and a complete SMR matrix of $h_{2a-b}(x)(x'Qx)^{\frac{b}{2}}$, built with respect to the same power vector $x^{\{m\}}$, i.e.*

$$h_{2a}(x) = x^{\{m\}'} W_1 x^{\{m\}}$$

$$h_{2a-b}(x)(x'Qx)^{\frac{b}{2}} = x^{\{m\}'} (W_2 + L(\alpha)) x^{\{m\}}.$$

Let us define the EVP

$$z^* = \inf_{z \in \mathbb{R}, \, \alpha \in \mathbb{R}^{\omega(n,m)}} z \tag{6.28}$$

$$\text{s.t. } zW_1 + W_2 + L(\alpha) > 0.$$

Then, $\hat{c}_{max}$ in (6.24) is given by

$$\hat{c}_{max} = z^{*\frac{2}{b}}. \tag{6.29}$$

Example 6.5. Let us consider problem (6.21) with

$$f(x) = -x_1^2 + x_1 x_2 + 2x_2^2 + 2x_1^4 - x_1^3 x_2 + 3x_1^2 x_2^2 - x_1 x_2^3 + x_2^4$$
$$Q = I_2.$$

Observe that $f(x)$ is as in (6.11) for

$$a = 2, \quad b = 2$$
$$h_{2a}(x) = -x_1^2 + x_1 x_2 + 2x_2^2$$
$$h_{2a+b}(x) = 2x_1^4 - x_1^3 x_2 + 3x_1^2 x_2^2 - x_1 x_2^3 + x_2^4.$$

From (6.29) we obtain the solution

$$\hat{c}_{max} = 0.5119.$$

Figure 6.6 shows the constraint $f(x) = 0$ and the circle $\mathscr{B}(I_2, \hat{c}_{max})$, from which we can deduce that the found upper bound is tight.

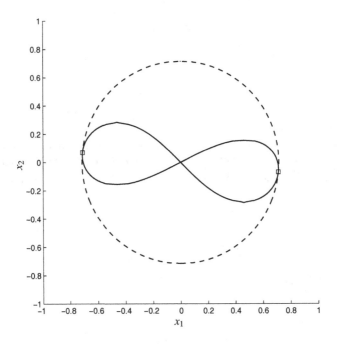

Fig. 6.6 Example 6.5: constraint $f(x) = 0$ (solid line) and $\mathscr{B}(I_2, \hat{c}_{max})$ (dashed line)

6.3 Conservatism Analysis

As shown in the previous section, a lower bound of c_{min} in (6.1) and an upper bound of c_{max} in (6.21) can be computed by solving LMI problems. Clearly, an important question concerns the possibility of establishing whether these bounds are tight.

6.3.1 a posteriori *Test for Tightness*

The following result allows one to devise a procedure for testing tightness of the lower bound $\hat{c}_{min}$ for the solution of the QDP (6.1), provided either by Theorem 6.1 or Theorems 6.2 and 6.3.

Theorem 6.7. *Let $\hat{c}_{min}$ be defined as in (6.10) and W^* be a maximal SMR matrix of $w(x; \hat{c}_{min})$. Then, $\hat{c}_{min} = c_{min}$ if and only if*

$$\exists \hat{x} \in \mathbb{R}_0^n : \hat{x}^{\{m\}} \in \ker(W^*). \tag{6.30}$$

Moreover, if (6.30) holds, an optimal point of problem (6.1) is given by

$$x_{min} = \gamma \hat{x} \sqrt{\frac{\hat{c}_{min}}{\hat{x}'Q\hat{x}}} \tag{6.31}$$

where $\gamma \in \{-1, 1\}$ is such that

$$f(x_{min}) = 0. \tag{6.32}$$

Lastly, if $w(x; c)$ is given by (6.8)–(6.9), then (6.31) holds with $\gamma = 1$.

Proof. (Necessity) Let us assume that $\hat{c}_{min} = c_{min}$ and let $x^* \in \mathbb{R}^n$ be an optimal point of (6.1). Then,

$$x^{*'}Qx^* = \hat{c}_{min}$$
$$f(x^*) = 0$$

and hence

$$w(x^*; \hat{c}_{min}) = 0.$$

Since $w(x; \hat{c}_{min})$ is SOS and W^* is a maximal SMR matrix of $w(x; \hat{c}_{min})$, it follows that W^* is positive semidefinite according to Definition 2.2. Hence, $\hat{x} = x^*$ satisfies (6.30). Moreover, (6.31) holds with $x_{min} = x^*$ and $\gamma = 1$.

(Sufficiency) Let us assume there exists $\hat{x} \in \mathbb{R}_0^n$ such that $\hat{x}^{\{m\}} \in \ker(W^*)$, and let us define

$$\tilde{x} = \hat{x} \sqrt{\frac{\hat{c}_{min}}{\hat{x}'Q\hat{x}}}.$$

We have by construction $\tilde{x}'Q\tilde{x} = \hat{c}_{min}$ and

$$w(\tilde{x}; \hat{c}_{min}) = 0.$$

If $w(x; c)$ is defined according to (6.6)–(6.7), this implies that

$$f(\tilde{x})f(-\tilde{x}) = 0.$$

By observing that x_{min} in (6.31) satisfies

$$x_{min} = \gamma \tilde{x}$$

one has that (6.32) hold with $\gamma \in \{-1, 1\}$, which implies that x_{min} is feasible. Being $\hat{c}_{min} \leq c_{min}$ by construction, we can conclude that $\hat{c}_{min} = c_{min}$ and (6.31) holds.

If $w(x;c)$ is defined according to (6.8)–(6.9), one has that $w(x;c) = f(x)$ for all $x \in \mathcal{B}(Q,c)$, and hence $f(\tilde{x}) = 0$. Therefore, the same reasoning adopted above implies that (6.31) holds with $\gamma = 1$. $\square$

Theorem 6.7 provides a necessary and sufficient condition for establishing tightness of the computed lower bound. This condition consists of the existence of a nonzero vector $\hat{x}$ such that $\hat{x}^{\{m\}}$ belongs to the linear space $\ker(W^*)$. Checking whether this condition is satisfied can be done by adopting the technique described in Section 1.9. Moreover, if such a vector $\hat{x}$ exists, then the simple scaling in (6.31) provides an optimal point for the problem (6.1).

A similar result holds for testing tightness of the upper bound $\hat{c}_{max}$ defined for the maximum QDP (6.21).

Let us show how the optimality test provided in Theorem 6.7 can be used in the previous examples.

Example 6.6. Let us consider the QDP in Example 6.1 with the lower bound $\hat{c}_{min} = 0.4901$ found in Example 6.2. We find that $\ker(W^*)$ is a linear space of dimension 1 for which a base is given by the vector

$$u = (0.6929, 0.5235, 0.3956, 0.2989)'.$$

Since the power vector $x^{\{m\}}$ has been chosen as

$$x^{\{m\}} = (x_1^3, x_1^2 x_2, x_1 x_2^2, x_2^3)'$$

we find that (6.30) holds with, e.g.,

$$\hat{x} = \begin{pmatrix} 1.0000 \\ 0.7556 \end{pmatrix}.$$

This implies that the found lower bound $\hat{c}_{min}$ is tight, by Theorem 6.7. Moreover, from (6.31) we have that

$$x_{min} = \gamma \begin{pmatrix} 0.5586 \\ 0.4220 \end{pmatrix}$$

is an optimal point of (6.1) for some $\gamma \in \{-1,1\}$. In order to select the value of γ, we evaluate $f(x_{min})$ and we find that (6.32) holds with $\gamma = -1$. In Figure 6.2 x_{min} is indicated by the square mark.

6.3.2 a priori *Conditions for Tightness*

It has been shown that the tightness of the lower bound $\hat{c}_{min}$ for the QDP (6.1) and upper bound $\hat{c}_{max}$ for (6.21) can be checked *a posteriori* through a numerical test. However, there exist structural conditions on the polynomial $f(x)$ ensuring *a priori* the tightness of the bounds. These conditions will be detailed in the following.

Theorem 6.8. *Let us suppose that* $(n, 2m) \in \mathcal{E}$ *for the QDP in Definition 6.1, with* $\mathcal{E}$ *given by (2.5). Then, the bounds* $\hat{c}_{min}$ *and* $\hat{c}_{max}$ *are tight, i.e.*

$$\hat{c}_{min} = c_{min}, \quad \hat{c}_{max} = c_{max}. \tag{6.33}$$

Proof. Let $(n, 2m) \in \mathcal{E}$. Then, from Theorem 2.4, for any given c one has that

$$w(x; c) > 0 \quad \forall x \in \mathbb{R}_0^n \iff \lambda(w(\cdot; c)) > 0$$

because $w(x; c) \in \Xi_{n,2m}$. Therefore, (6.10) is equivalent to (6.4), and (6.24) is equivalent to (6.23). □

Example 6.7. For all the QDPs considered in Examples 6.1–6.5, one has $(n, 2m) \in \mathcal{E}$, which implies that the bounds achievable by the described techniques are known *a priori* to be tight, by Theorem 6.8.

6.3.3 An Example of Nontight Lower Bound

In the following, an example is presented for which the lower bound $\hat{c}_{min}$ does not coincide with the solution of the QDP. Let us consider problem (6.1) with

$$f(x) = 1 + \left(10 g_{Mot}(x) - (x_1^2 + x_2^2 + x_3^2)^3 \right)$$
$$Q = I_3$$

where $g_{Mot}(x) \in \Xi_{3,6}$ is the Motzkin form in (2.1). It can be easily verified that the resulting form $w(x; c)$ is given by

$$w(x; c) = 10 g_{Mot}(x) + \frac{1 - c^3}{c^3} \|x\|^6.$$

Since $g_{Mot}(x)$ is positive semidefinite, from (6.4) one has $c_{min} \geq 1$. Moreover, $f(x) = 0$ for $x = \frac{1}{\sqrt{3}}(1, 1, 1)'$. Therefore, the solution of (6.1) is $c_{min} = 1$. However,

$$w(x; 1) = 10 g_{Mot}(x)$$

which implies that $w(x; 1)$ is PNS because $g_{Mot}(x)$ is itself PNS. Hence, $\hat{c}_{min} < c_{min}$ from Theorem 6.1. Indeed, it can be checked that $\hat{c}_{min} = 0.9851$.

Let us observe that the lower bound provided by the described approach can be made tight for the considered QDP by suitably increasing the degree of the polynomial $f(x)$. In fact, as explained in Theorem 2.1, each PNS form can be expressed as the ratio of two SOS forms, which means that non-conservatism can be achieved by multiplying $f(x)$ by a suitable SOS form. Indeed, let us define the equivalent polynomial constraint

$$\tilde{f}(x) = f(x)(x_1^2 + x_2^2 + x_3^2 + 1).$$

In this case, we find that $\hat{c}_{min} = 1$, i.e. the lower bound is tight. This is due to the fact that the corresponding form $\tilde{w}(x;c)$ satisfies

$$\tilde{w}(x;1) = 20 g_{Mot}(x)(x_1^2 + x_2^2 + x_3^2)$$

and it turns out to be an SOS form.

6.4 l_2 Parametric Stability Margin

The parametric stability margin is an important measure of the robustness of control systems affected by time-invariant parametric uncertainty. Roughly speaking, such margin identifies the largest region of a given shape in the parameter space, where stability of the nominal control system is preserved. For linear systems it turns out that, when such a shape is defined in terms of an euclidean norm, the computation of the parametric stability margin amounts to solving a QDP.

6.4.1 Continuous-time Systems

Let us consider the continuous-time time-invariant uncertain system

$$\dot{x}(t) = A(p)x(t) \tag{6.34}$$

where $x(t) \in \mathbb{R}^n$ is the state vector, $p \in \mathbb{R}^q$ is an uncertain parameter vector, and $A(p)$ has the form

$$A(p) = A_0 + \sum_{i=1}^{q} p_i A_i. \tag{6.35}$$

The matrix $A_0 \in \mathbb{R}^{n \times n}$ represents the nominal control system and is assumed to be Hurwitz, while the matrices $A_i \in \mathbb{R}^{n \times n}$, $i = 1, \ldots, q$, model the uncertainty structure. We introduce the following definition.

Definition 6.2 (l_2 Parametric Stability Margin). Let us define

$$\gamma^{PQ} = \sup \ \{\gamma \in \mathbb{R} : A(p) \text{ is Hurwitz } \forall p : \|p\| \le \gamma\}. \tag{6.36}$$

Then, γ^{PQ} is said l_2 *parametric stability margin* of the system (6.34)–(6.35).

In order to compute γ^{PQ}, let us observe that

$$0 \in \mathrm{spc}(A(p)) \quad \Longleftrightarrow \quad \det(A(p)) = 0.$$

Moreover,

$$A(p) \text{ is Hurwitz } \Rightarrow \hbar_{n-1}(A(p)) \neq 0$$

and

$$\exists \omega \in \mathbb{R}_0 : j\omega \in \text{spc}(A(p)) \Rightarrow \hbar_{n-1}(A(p)) = 0$$

where $\hbar_{n-1}(A(p))$ is the $(n-1)$-th Hurwitz determinant of the matrix $A(p)$ (see Appendix A.2 for details). Hence, it is straightforward to verify that γ^{PQ} can be computed as follows

$$\gamma^{PQ} = \min \{\sqrt{\gamma_I}, \sqrt{\gamma_{II}}\} \tag{6.37}$$

where

$$\gamma_I = \inf_{p \in \mathbb{R}^q} \|p\|^2 \tag{6.38}$$
$$\text{s.t. } \det(A(p)) = 0$$

and

$$\gamma_{II} = \inf_{p \in \mathbb{R}^q} \|p\|^2 \tag{6.39}$$
$$\text{s.t. } \hbar_{n-1}(A(p)) = 0.$$

Problem (6.38) (resp. (6.39)) is a QDP, once that

$$\begin{cases} c_{min} = \gamma_I & (\text{resp. } \gamma_{II}) \\ x = p \\ Q = I_q \\ f(x) = \det(A(p)) & (\text{resp. } \hbar_{n-1}(A(p))) \\ n = q \\ m = m_I & (\text{resp. } m_{II}) \end{cases} \tag{6.40}$$

where m_I (resp. m_{II}) denotes the degree of $\det(A(p))$ (resp. $\hbar_{n-1}(A(p))$). In particular, one has

$$m_I \leq n \tag{6.41}$$
$$m_{II} \leq \frac{n(n-1)}{2}. \tag{6.42}$$

Example 6.8. Let us consider the problem of computing γ^{PQ} for the matrix

$$A(p) = \begin{pmatrix} -1 & p_1 & 1 \\ 1 & p_1 - 2 & p_2 \\ 0 & 1 & p_2 - 1 \end{pmatrix}.$$

Let us consider first γ_I. We have that the constraint of the QDP is defined by

$$\det(A(p)) = -1 + 2p_1 + 3p_2 - 2p_1 p_2.$$

From (6.10) we find the lower bound $\hat{\gamma}_I = 0.0897$, shown in Figure 6.7. Observe that this lower bound is tight, and it must be according to Theorem 6.8. Theorem

6.7 provides the optimal point

$$p_{min} = \begin{pmatrix} 0.1429 \\ 0.2631 \end{pmatrix}.$$

Now, let us consider γ_{II}. We have that the constraint of the QDP is defined by the polynomial

$$\hbar_{n-1}(A(p)) = -19 + 15p_1 + 18p_2 - 3p_1^2 - 9p_1p_2 - 4p_2^2 + p_1^2p_2 + p_1p_2^2.$$

From (6.10) we find the lower bound $\hat{\gamma}_{II} = 1.9586$, shown in Figure 6.8. Also this lower bound is tight in accordance with Theorem 6.8. The optimal point provided by Theorem 6.7 is

$$p_{min} = \begin{pmatrix} 0.8333 \\ 1.1244 \end{pmatrix}.$$

Lastly, from (6.37) we conclude that $\gamma^{PQ} = \gamma_I = 0.0897$.

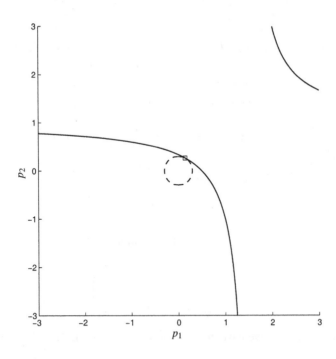

Fig. 6.7 Example 6.8: constraint $\det(A(p)) = 0$ (solid line) and global minimum $\hat{\gamma}_I$ (dashed line)

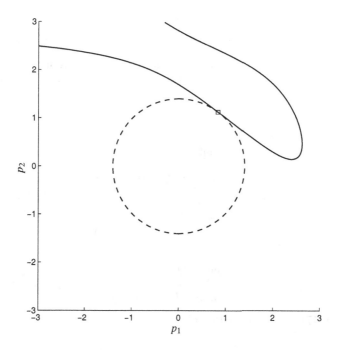

Fig. 6.8 Example 6.8: constraint $\hbar_{n-1}(A(p)) = 0$ (solid line) and global minimum $\hat{\gamma}_{II}$ (dashed line)

6.4.2 Discrete-time Systems

Let us consider the discrete-time time-invariant uncertain system

$$x(t+1) = A(p)x(t) \qquad (6.43)$$

where $t \in \mathbb{N}$, $x(t) \in \mathbb{R}^n$ is the state vector, and $p \in \mathbb{R}^q$ is an uncertain parameter vector. The matrix $A(p)$ has the form in (6.35) for some matrices $A_0, \dots, A_q \in \mathbb{R}^{n \times n}$. The l_2 parametric stability margin for the system (6.43) is defined as follows.

Definition 6.3 (l_2 Parametric Stability Margin for Discrete-time Systems). Let us define

$$\gamma^{PQ} = \sup \ \{ \gamma \in \mathbb{R} : \ A(p) \text{ is Schur}, \ \forall p : \ \|p\| \le \gamma \}. \qquad (6.44)$$

Then, γ^{PQ} is said l_2 *parametric stability margin* of the system (6.43).

In order to compute γ^{PQ}, let us observe that

$$\pm 1 \in \text{spc}(A(p)) \iff \det(I_n \mp A(p)) = 0.$$

Moreover,

$$A(p) \text{ is Schur} \Rightarrow \det(I_d - Q(A(p))) \ne 0 \qquad (6.45)$$

and
$$\exists \omega \in (0, \pi) : \quad e^{j\omega} \in \text{spc}(A(p)) \quad \Rightarrow \quad \det(I_d - Q(A(p))) = 0$$

where $d = \frac{1}{2}n(n-1)$ and $Q(A(p))$ is the matrix defined according to (A.5) in Appendix A.2. Hence, it is straightforward to verify that γ^{PQ} can be computed as follows:

$$\gamma^{PQ} = \min \{\sqrt{\gamma_I}, \sqrt{\gamma_{II}}, \sqrt{\gamma_{III}}\} \tag{6.46}$$

where

$$\gamma_I = \inf_{p \in \mathbb{R}^q} \|p\|^2$$
$$\text{s.t.} \quad \det(I_n - A(p)) = 0 \tag{6.47}$$

$$\gamma_{II} = \inf_{p \in \mathbb{R}^q} \|p\|^2$$
$$\text{s.t.} \quad \det(I_n + A(p)) = 0 \tag{6.48}$$

and

$$\gamma_{III} = \inf_{p \in \mathbb{R}^q} \|p\|^2$$
$$\text{s.t.} \quad \det(I_d - Q(A(p))) = 0. \tag{6.49}$$

Problems (6.47)–(6.49) are QDPs, once that

$$\begin{cases} c_{min} = \gamma_I & \text{(resp. } \gamma_{II}) & \text{(resp. } \gamma_{III}) \\ x = p \\ Q = I_n \\ f(x) = \det(I_n - A(p)) & \text{(resp. } \det(I_n + A(p))) & \text{(resp. } \det(I_d - Q(A(p)))) \\ n = q \\ m = m_I & \text{(resp. } m_{II}) & \text{(resp. } m_{III}) \end{cases}$$

where m_I, m_{II} and m_{III} denote the degrees of $\det(I_n - A(p))$, $\det(I_n + A(p))$ and $\det(I_d - Q(A(p)))$, respectively. In particular, one has

$$m_I \le n$$
$$m_{II} \le n$$
$$m_{III} \le 2d.$$

6.5 Notes and References

The solution of QDPs via LMI feasibility tests was first proposed in [51, 50]. The main results reported in this chapter have been presented in [52, 36].

A number of problems relevant to system analysis and design can be cast as QDPs. Just to mention a few: the estimation of the domain of attraction of equilibria of nonlinear systems via quadratic Lyapunov functions [67]; the D-stability of real matrices [76], which plays a key role in the analysis of singularly perturbed systems [84]; the computation of the region of validity of optimal linear $\mathscr{H}_\infty$ controllers for nonlinear systems [141]; the characterization of the frequency plots of an ellipsoidal family of rational functions [31].

The l_2 stability margin of a control system affected by parametric uncertainty is a key problem in robust control [5]. The approach based on Hurwitz determinants, exploited in Section 4.5, is quite popular in the literature, see e.g. [136, 143] and references therein.

Other LMI-based relaxations for the solution of QDPs, and of more general classes of polynomial optimization problems, have been formulated by exploiting Stengle's Positivstellensatz (see e.g. [133, 106, 137, 89]), the theory of moments [85, 74], and the slack variable approach [110]. An equivalence result between two different relaxations for QDPs has been given in [66].

Appendix A
Basic Tools

A.1 LMI Optimizations

We provide some basic notions about the convex optimization problems considered in the book. The interested reader is referred to [14, 15] for an extensive treatment.

Definition A.1 (Linear Matrix Function). The function $A(X) : \mathbb{R}^{n \times m} \to \mathbb{R}^{p \times q}$ is a *linear matrix function* if

$$A(\alpha X + \beta Y) = \alpha A(X) + \beta A(Y) \quad \forall X, Y \in \mathbb{R}^{n \times m} \ \forall \alpha, \beta \in \mathbb{R}. \qquad (A.1)$$

Moreover, if $A(X) = A(X)'$, then $A(X)$ is said a *symmetric linear matrix function*.

Definition A.2 (LMI). Let $A(X) : \mathbb{R}^{n \times m} \to \mathbb{S}^p$ be a symmetric linear matrix function, and let $A_0 \in \mathbb{S}^p$ be a constant symmetric matrix. Then, the constraint

$$A(X) + A_0 * 0$$

is an *LMI* for any symbol $*$ in the set $\{>, <, \geq, \leq\}$.

Definition A.3 (LMI Feasibility Set). For $i = 1, \ldots, k$ let $A_i(X) : \mathbb{R}^{n \times m} \to \mathbb{S}^{p_i}$ be a symmetric linear matrix function, and let $A_{0,i} \in \mathbb{S}^{p_i}$ be a constant symmetric matrix. Let us define the set of LMIs

$$A_i(X) + A_{0,i} * 0 \ \forall i = 1, \ldots, k. \qquad (A.2)$$

Then, the set

$$\{X \in \mathbb{R}^{n \times m} : A_i(X) + A_{0,i} * 0 \ \forall i = 1, \ldots, k\}$$

is the *feasibility set* of the set of LMIs (A.2).

The feasibility set of a set of LMIs is a convex set [14].

Definition A.4 (LMI Feasibility Test). For $i = 1, \ldots, k$ let $A_i(X) : \mathbb{R}^{n \times m} \to \mathbb{S}^{p_i}$ be a symmetric linear matrix function, and let $A_{0,i} \in \mathbb{S}^{p_i}$ be a constant symmetric matrix. The problem of establishing whether

$$\exists X \in \mathbb{R}^{n \times m} : A_i(X) + A_{0,i} * 0 \ \forall i = 1, \ldots, k \tag{A.3}$$

is an *LMI feasibility test*.

Solving an LMI feasibility test amounts to solving a convex optimization problem [14].

Definition A.5 (EVP). For $i = 1, \ldots, k$ let $A_i(X) : \mathbb{R}^{n \times m} \to \mathbb{S}^{p_i}$ be a symmetric linear matrix function, and let $A_{0,i} \in \mathbb{S}^{p_i}$ be a constant symmetric matrix. Let $c(X) : \mathbb{R}^{n \times m} \to \mathbb{R}$ be a linear function. The optimization problem

$$\inf_X \ c(X)$$
$$\text{s.t. } A_i(X) + A_{0,i} * 0 \ \forall i = 1, \ldots, k$$

is an *EVP*.

An EVP is a convex optimization problem [14]. EVPs are also known as semidefinite programs [142].

Definition A.6 (GEVP). For $i = 1, \ldots, k$ let $A_i(X), B_i(X) : \mathbb{R}^{n \times m} \to \mathbb{S}^{p_i}$ and $C_i(X) : \mathbb{R}^{n \times m} \to \mathbb{S}^{q_i}$ be symmetric linear matrix functions, and let $A_{0,i}, B_{0,i} \in \mathbb{S}^{p_i}$ and $C_{0,i} \in \mathbb{S}^{q_i}$ be constant symmetric matrices. The optimization problem

$$\inf_{t,X} \ t$$
$$\text{s.t. } \begin{cases} t \left(B_i(X) + B_{0,i} \right) - A_i(X) - A_{0,i} > 0 \\ B_i(X) + B_{0,i} > 0 \\ C_i(X) + C_{0,i} * 0 \\ i = 1, \ldots, k \end{cases}$$

is a *GEVP*.

A GEVP is a quasiconvex optimization problem [14].

A.2 Hurwitz/Schur Matrices

Definition A.7 (Hurwitz Matrix). A matrix $A \in \mathbb{R}^{n \times n}$ is said *Hurwitz* if all eigenvalues of A lie on the left open complex half-plane, i.e.

$$\text{re}(\lambda) < 0 \ \forall \lambda \in \text{spc}(A).$$

Definition A.8 (Schur Matrix). A matrix $A \in \mathbb{R}^{n \times n}$ is said *Schur* if all eigenvalues of A lie inside the open unit disc of the complex plane, i.e.

$$|\lambda| < 1 \quad \forall \lambda \in \mathrm{spc}(A).$$

Definition A.9 (Hurwitz Determinant). Let us consider $A \in \mathbb{R}^{n \times n}$. The *Hurwitz determinants* of A are defined as

$$\hbar_1(A) = p_{n-1}$$

$$\hbar_2(A) = \det \begin{pmatrix} p_{n-1} & p_{n-3} \\ 1 & p_{n-2} \end{pmatrix}$$

$$\hbar_3(A) = \det \begin{pmatrix} p_{n-1} & p_{n-3} & p_{n-5} \\ 1 & p_{n-2} & p_{n-4} \\ 0 & p_{n-1} & p_{n-3} \end{pmatrix}$$

$$\vdots$$

$$\hbar_n(A) = \det \begin{pmatrix} p_{n-1} & p_{n-3} & p_{n-5} & \cdots & 0 \\ 1 & p_{n-2} & p_{n-4} & \cdots & 0 \\ 0 & p_{n-1} & p_{n-3} & \cdots & 0 \\ 0 & 1 & p_{n-2} & \cdots & 0 \\ \vdots & & & & \\ 0 & & \cdots & & p_0 \end{pmatrix}$$

where $p_0, \ldots, p_{n-1} \in \mathbb{R}$ are the coefficients of the characteristic polynomial $p(x)$ of A, which is given by

$$p(x) = \det(xI_n - A) = x^n + \sum_{i=0}^{n-1} p_i x^i.$$

In particular, $\hbar_i(A)$ is the *i-th Hurwitz determinant* of A.

As explained in [64], it turns out that $\hbar_{n-1}(A)$ is a polynomial in the elements of A of degree

$$d = \frac{1}{2} n(n-1).$$

Moreover,

$$\hbar_{n-1}(A) = (-1)^d \prod_{\substack{i=1,\ldots,n-1 \\ j=i+1,\ldots,n}} (\lambda_i(A) + \lambda_j(A)) \tag{A.4}$$

where $\lambda_1(A), \ldots, \lambda_n(A) \in \mathbb{C}$ are the eigenvalues of A.

Let us consider $A \in \mathbb{R}^{n \times n}$. Let us define the matrix

$$Q(A) = \begin{pmatrix} q(p_1,p_1) & q(p_1,p_2) & \cdots & q(p_1,p_d) \\ q(p_2,p_1) & q(p_2,p_2) & \cdots & q(p_2,p_d) \\ \vdots & \vdots & \ddots & \vdots \\ q(p_d,p_1) & q(p_d,p_2) & \cdots & q(p_d,p_d) \end{pmatrix} \qquad (A.5)$$

where $p_1, p_2, \ldots, p_d$ are the pairs

$$(1,2),(1,3),\ldots,(1,n),(2,3),(2,4),\ldots,(2,n),(3,4),\ldots,(n-1,n),$$

$q(p_i, p_j)$ is the function

$$q(p_i,p_j) = \det \begin{pmatrix} A_{x_1,y_1} & A_{x_1,y_2} \\ A_{x_2,y_1} & A_{x_2,y_2} \end{pmatrix}, \quad p_i = (x_1,x_2), \ p_j = (y_1,y_2),$$

and

$$d = \frac{1}{2}n(n-1).$$

It turns out

$$\mathrm{spc}(Q(A)) = \{z = \lambda_i(A)\lambda_j(A) : i = 1,\ldots,n-1, \ j = i+1,\ldots,n\}$$

where $\lambda_1(A),\ldots,\lambda_n(A) \in \mathbb{C}$ are the eigenvalues of A (see e.g. [4]).

Appendix B
SMR Algorithms

Let $\mathrm{ind}(a(x), x^{\{m\}})$ be the position of the monomial $a(x)$ in $x^{\{m\}}$. Tables B.1–B.4 provide algorithms for the construction of the matrices involved in the complete SMR of forms (1.17) and the complete SMR of matrix forms (1.41).

Table B.1 Algorithm 1: computation of an SMR matrix H in (1.17)

Step	Instruction
1	choose $x^{\{m\}}$ and $x^{\{2m\}}$ as in (1.7)
2	let $g \in \mathbb{R}^{\sigma(n,2m)}$ be such that $h(x) = g'x^{\{2m\}}$
3	set $H = 0_{\sigma(n,m) \times \sigma(n,m)}$
4	for $i = 1, \ldots, \sigma(n,m)$ and $j = i, \ldots, \sigma(n,m)$
5	set $a = \mathrm{ind}\left((x^{\{m\}})_i(x^{\{m\}})_j, x^{\{2m\}} \right)$ and $H_{i,j} = H_{i,j} + g_a$
6	endfor
7	set $H = 0.5\,\mathrm{he}(H)$

Table B.2 Algorithm 2: computation of an SMR matrix H for matrix forms in (1.41)

Step	Instruction
1	choose $x^{\{m\}}$ and $x^{\{2m\}}$ as in (1.7)
2	set $H = 0_{r\sigma(n,m) \times r\sigma(n,m)}$
3	for $i = 1, \ldots, r$ and $j = 1, \ldots, r$
4	let G be an SMR matrix of $M_{i,j}(x)$ with respect to $x^{\{m\}}$ (e.g., built via Algorithm 1)
5	for $k = 1, \ldots, \sigma(n,m)$ and $l = 1, \ldots, \sigma(n,m)$
6	set $a = r(k-1)+i$ and $b = r(l-1)+j$
7	set $H_{a,b} = H_{a,b} + G_{k,l}$
8	endfor
9	endfor
10	set $H = 0.5\,\mathrm{he}(H)$

Table B.3 Algorithm 3: computation of an SMR parametrization $L(\alpha)$ in (1.17)

Step	Instruction
1	choose $x^{\{m\}}$ and $x^{\{2m\}}$ as in (1.7)
2	set $A = 0_{\sigma(n,2m) \times 3}$ and $b = 0$ and define the variable $\alpha \in \mathbb{R}^{\omega(n,m)}$
3	for $i = 1,\ldots,\sigma(n,m)$ and $j = i,\ldots,\sigma(n,m)$
4	set $a = \mathrm{ind}\left((x^{\{m\}})_i (x^{\{m\}})_j, x^{\{2m\}} \right)$ and $A_{a,1} = A_{a,1} + 1$
5	if $A_{a,1} = 1$
6	set $A_{a,2} = i$ and $A_{a,3} = j$
7	else
8	set $b = b+1$ and $G = 0_{\sigma(n,m) \times \sigma(n,m)}$ and $G_{i,j} = 1$
9	set $k = A_{a,2}$ and $l = A_{a,3}$ and $G_{k,l} = G_{k,l} - 1$
10	set $L(\alpha) = L(\alpha) + \alpha_b G$
11	endif
12	endfor
13	set $L(\alpha) = 0.5 \,\mathrm{he}(L(\alpha))$

Table B.4 Algorithm 4: computation of an SMR parametrization $L(\alpha)$ for matrix forms in (1.41)

Step	Instruction
1	choose $x^{\{m\}}$ and $x^{\{2m\}}$ as in (1.7)
2	set $A = 0_{\sigma(n,2m)\sigma(r,2) \times 3}$ and $b = 0$ and define the variable $\alpha \in \mathbb{R}^{\omega(n,m,r)}$
3	for $i = 1,\ldots,\sigma(n,m)$ and $j = i,\ldots,r$
4	set $c = r(i-1) + j$
5	for $k = i,\ldots,\sigma(n,m)$ and $l = \max\{1, j+r(i-k)\},\ldots,r$
6	set $d = r(k-1) + l$ and $f = \mathrm{ind}\left((x^{\{m\}})_i (x^{\{m\}})_k, x^{\{2m\}} \right)$
7	set $g = \mathrm{ind}\left(y_j y_l, y^{\{2\}} \right)$ and $a = f\sigma(r,2) + g$ and $A_{a,1} = A_{a,1} + 1$
8	if $A_{a,1} = 1$
9	set $A_{a,2} = c$ and $A_{a,3} = d$
10	else
11	set $b = b+1$ and $G = 0_{r\sigma(n,m) \times r\sigma(n,m)}$ and $G_{c,d} = 1$
12	set $h = A_{a,2}$ and $p = A_{a,3}$ and $G_{h,p} = G_{h,p} - 1$
13	set $L(\alpha) = L(\alpha) + \alpha_b G$
14	endif
15	endfor
16	endfor
17	set $L(\alpha) = 0.5 \,\mathrm{he}(L(\alpha))$

References

1. Almeida, H.L.S., Bhaya, A., Falcao, D.M., Kaszkurewicz, E.: A team algorithm for robust stability analysis and control design of uncertain time-varying linear systems using piecewise quadratics Lyapunov functions. Int. Journal of Robust and Nonlinear Control 11, 357–371 (2001)
2. Altafini, C.: Homogeneous polynomial forms for simultaneous stabilizability of families of linear control systems: a tensor product approach. IEEE Transactions on Automatic Control 51(9), 1566–1571 (2006)
3. Barmish, B.R.: New Tools for Robustness of Linear Systems. Mcmillan Publishing Company, New York (1994)
4. Bellman, R.: Introduction to Matrix Analysis. McGraw-Hill, New York (1974)
5. Bhattacharyya, S.P., Chapellat, H., Keel, L.H.: Robust Control: The Parametric Approach. Prentice Hall, NJ (1995)
6. Blanchini, F.: Nonquadratic Lyapunov functions for robust control. Automatica 31, 451–461 (1995)
7. Blanchini, F., Miani, S.: On the transient estimate for linear systems with time-varying uncertain parameters. IEEE Trans. on Circuits and Systems I 43, 592–596 (1996)
8. Blanchini, F., Miani, S.: A universal class of smooth functions for robust control. IEEE Trans. on Automatic Control 44(3), 641–647 (1999)
9. Blekherman, G.: There are significantly more nonnegative polynomials than sums of squares. Israel Journal of Mathematics 153(1), 355–380 (2006)
10. Bliman, P.-A.: Nonconservative LMI approach to robust stability for systems with uncertain scalar parameters. In: IEEE Conf. on Decision and Control, Las Vegas, Nevada, pp. 305–310 (2002)
11. Bliman, P.-A.: A convex approach to robust stability for linear systems with uncertain scalar parameters. SIAM Journal on Control and Optimization 42(6), 2016–2042 (2004)
12. Bliman, P.-A., Oliveira, R.C.L.F., Montagner, V.F., Peres, P.L.D.: Existence of homogeneous polynomial solutions for parameter-dependent linear matrix inequalities with parameters in the simplex. In: IEEE Conf. on Decision and Control, San Diego, CA, pp. 1486–1491 (2006)
13. Bose, N.K.: Applied Multidimensional Systems Theory. Van Rostrand Reinhold, New York (1982)

14. Boyd, S., Ghaoui, L.E., Feron, E., Balakrishnan, V.: Linear Matrix Inequalities in System and Control Theory. SIAM, Philadelphia (1994)
15. Boyd, S., Vandenberghe, L.: Convex Optimization. Cambridge University Press, Cambridge (2004)
16. Boyd, S., Yang, Q.: Structured and simultaneous Lyapunov functions for system stability problems. Int. Journal of Control 49(6), 2215–2240 (1989)
17. Brayton, R.K., Tong, C.H.: Stability of dynamical systems: a constructive approach. IEEE Trans. on Circuits and Systems 26, 224–234 (1979)
18. Brockett, R.W.: Finite dimensional linear systems. John Wiley & Sons, New York (1970)
19. Brockett, R.W.: Lie algebra and Lie groups in control theory. In: Mayne, D.Q., Brockett, R.W. (eds.) Geometric Methods in Systems Theory, pp. 43–82. Reidel, Dordrecht (1973)
20. Chen, Y.H., Chen, J.S.: Robust control of uncertain systems with time-varying uncertainty: a computer-aided setting. In: Skowronski, J.M., Flashner, H., Guttalu, R.S. (eds.) Positive Polynomials in Control. Lecture Notes in Control and Information Sciences, vol. 151, pp. 97–114. Springer, New York (1991)
21. Chesi, G.: Computing output feedback controllers to enlarge the domain of attraction in polynomial systems. IEEE Trans. on Automatic Control 49(10), 1846–1850 (2004)
22. Chesi, G.: Estimating the domain of attraction for uncertain polynomial systems. Automatica 40(11), 1981–1986 (2004)
23. Chesi, G.: Establishing stability and instability of matrix hypercubes. Systems and Control Letters 54(4), 381–388 (2005)
24. Chesi, G.: Establishing tightness in robust H-infinity analysis via homogeneous parameter-dependent Lyapunov functions. Automatica 43(11), 1992–1995 (2007)
25. Chesi, G.: On the gap between positive polynomials and SOS of polynomials. IEEE Trans. on Automatic Control 52(6), 1066–1072 (2007)
26. Chesi, G.: On the non-conservatism of a novel LMI relaxation for robust analysis of polytopic systems. Automatica 44(11), 2973–2976 (2008)
27. Chesi, G.: Tightness conditions for semidefinite relaxations of forms minimizations. IEEE Trans. on Circuits and Systems II 55(12), 1299–1303 (2008)
28. Chesi, G.: Camera displacement via constrained minimization of the algebraic error. IEEE Trans. on Pattern Analysis and Machine Intelligence 31(2), 370–375 (2009)
29. Chesi, G.: On the minimum stable commutation time for switching nonlinear systems. IEEE Trans. on Automatic Control 54(6), 1284–1289 (2009)
30. Chesi, G.: Visual servoing path-planning via homogeneous forms and LMI optimizations. IEEE Trans. on Robotics 25(2), 281–291 (2009)
31. Chesi, G., Garulli, A., Tesi, A., Vicino, A.: A convex approach to the characterization of the frequency response of ellipsoidal plants. Automatica 38(2), 249–259 (2002)
32. Chesi, G., Garulli, A., Tesi, A., Vicino, A.: LMI-based construction of homogeneous Lyapunov functions for systems with structured uncertainties. In: 41st IEEE Conf. on Decision and Control, Las Vegas, Nevada, pp. 281–286 (2002)
33. Chesi, G., Garulli, A., Tesi, A., Vicino, A.: Characterizing the solution set of polynomial systems in terms of homogeneous forms: an LMI approach. Int. Journal of Robust and Nonlinear Control 13(13), 1239–1257 (2003)
34. Chesi, G., Garulli, A., Tesi, A., Vicino, A.: Homogeneous Lyapunov functions for systems with structured uncertainties. Automatica 39(6), 1027–1035 (2003)
35. Chesi, G., Garulli, A., Tesi, A., Vicino, A.: Robust stability for polytopic systems via polynomially parameter-dependent Lyapunov functions. In: 42nd IEEE Conf. on Decision and Control, Maui, Hawaii, pp. 4670–4675 (2003)

36. Chesi, G., Garulli, A., Tesi, A., Vicino, A.: Solving quadratic distance problems: an LMI-based approach. IEEE Trans. on Automatic Control 48(2), 200–212 (2003)
37. Chesi, G., Garulli, A., Tesi, A., Vicino, A.: Parameter-dependent homogeneous Lyapunov functions for robust stability of linear time-varying systems. In: 43rd IEEE Conf. on Decision and Control, Paradise Island, Bahamas, pp. 4095–4100 (2004)
38. Chesi, G., Garulli, A., Tesi, A., Vicino, A.: Robust analysis of LFR systems through homogeneous polynomial Lyapunov functions. IEEE Trans. on Automatic Control 49(7), 1211–1216 (2004)
39. Chesi, G., Garulli, A., Tesi, A., Vicino, A.: LMI-based computation of optimal quadratic Lyapunov functions for odd polynomial systems. Int. Journal of Robust and Nonlinear Control 15(1), 35–49 (2005)
40. Chesi, G., Garulli, A., Tesi, A., Vicino, A.: An LMI-based technique for robust stability analysis of linear systems with polynomial parametric uncertainties. In: Henrion, D., Garulli, A. (eds.) Positive Polynomials in Control. Lecture Notes in Control and Information Sciences, vol. 312. Springer, London (2005)
41. Chesi, G., Garulli, A., Tesi, A., Vicino, A.: Polynomially parameter-dependent Lyapunov functions for robust $\mathcal{H}_\infty$ performance analysis. In: 16th IFAC World Congress on Automatic Control, Prague, Czech Republic (2005)
42. Chesi, G., Garulli, A., Tesi, A., Vicino, A.: Polynomially parameter-dependent Lyapunov functions for robust stability of polytopic systems: an LMI approach. IEEE Trans. on Automatic Control 50(3), 365–370 (2005)
43. Chesi, G., Garulli, A., Tesi, A., Vicino, A.: Robust stability of time-varying polytopic systems via parameter-dependent homogeneous Lyapunov functions. Automatica 43(2), 309–316 (2007)
44. Chesi, G., Garulli, A., Vicino, A., Cipolla, R.: Estimating the fundamental matrix via constrained least-squares: a convex approach. IEEE Trans. on Pattern Analysis and Machine Intelligence 24(3), 397–401 (2002)
45. Chesi, G., Henrion, D. (eds.): Special Issue on Positive Polynomials in Control. IEEE Trans. on Automatic Control, vol. 54 (2009)
46. Chesi, G., Hung, Y.S.: Image noise induced errors in camera positioning. IEEE Trans. on Pattern Analysis and Machine Intelligence 29(8), 1476–1480 (2007)
47. Chesi, G., Hung, Y.S.: Analysis and synthesis of nonlinear systems with uncertain initial conditions. IEEE Trans. on Automatic Control 53(5), 1262–1267 (2008)
48. Chesi, G., Hung, Y.S.: Establishing convexity of polynomial Lyapunov functions and their sublevel sets. IEEE Trans. on Automatic Control 53(10), 2431–2436 (2008)
49. Chesi, G., Hung, Y.S.: Stability analysis of uncertain genetic SUM regulatory networks. Automatica 44(9), 2298–2305 (2008)
50. Chesi, G., Tesi, A., Vicino, A., Genesio, R.: A convex approach to a class of minimum norm problems. In: Garulli, A., Tesi, A., Vicino, A. (eds.) Robustness in Identification and Control, pp. 359–372. Springer, London (1999)
51. Chesi, G., Tesi, A., Vicino, A., Genesio, R.: On convexification of some minimum distance problems. In: 5th European Control Conf., Karlsruhe, Germany (1999)
52. Chesi, G., Tesi, A., Vicino, A., Genesio, R.: An LMI approach to constrained optimization with homogeneous forms. Systems and Control Letters 42(1), 11–19 (2001)
53. Choi, M., Lam, T., Reznick, B.: Sums of squares of real polynomials. In: Symposia in Pure Mathematics, pp. 103–126 (1995)
54. Cobb, J.D., DeMarco, C.L.: The minimal dimension of stable faces to guarantee stability of a matrix polytope. IEEE Trans. on Automatic Control 34(9), 990–992 (1989)
55. Curto, R.E., Fialkow, L.A.: The truncated complex K-moment problem. Transactions of the American Mathematical Society 352(6), 2825–2855 (2000)

56. Dasgupta, S., Chockalingam, G., Anderson, B.D.O., Fu, M.: Lyapunov functions for uncertain systems with applications to the stability of time varying systems. IEEE Trans. on Automatic Control 41, 93–106 (1994)

57. de Klerk, E., Laurent, M., Parrilo, P.: On the equivalence of algebraic approaches to the minimization of forms on the simplex. In: Henrion, D., Garulli, A. (eds.) Positive polynomials in Control. Lecture Notes in Control and Information Sciences, pp. 121–132. Springer, Heidelberg (2005)

58. de Oliveira, M.C., Bernussou, J., Geromel, J.C.: A new discrete-time robust stability condition. Systems and Control Letters 37, 261–265 (1999)

59. de Oliveira, P.J., Oliveira, R.C.L.F., Leite, V.J.S., Montagner, V.F., Peres, P.L.D.: $\mathcal{H}_\infty$ guaranteed cost computation by means of parameter-dependent Lyapunov functions. Automatica 40(6), 1053–1061 (2004)

60. Ebihara, Y., Onishi, Y., Hagiwara, T.: Robust performance analysis of uncertain LTI systems: Dual LMI approach and verifications for exactness. In: Chesi, G., Henrion, D. (eds.) Special Issue on Positive Polynomials in Control. IEEE Trans. on Automatic Control, vol. 54, pp. 938–951 (2009)

61. El-Samad, H., Prajna, S., Papachristodoulou, A., Doyle, J., Khammash, M.: Advanced methods and algorithms for biological networks analysis. Proceedings of the IEEE 94(4), 832–853 (2006)

62. Fujisaki, Y., Sakuwa, R.: Estimation of asymptotic stability regions via homogeneous polynomial Lyapunov functions. International Journal of Control 79(6), 617–623 (2006)

63. Fukumoto, H., Fujisaki, Y.: Exact robust $\mathcal{H}_2$ performance analysis for linear single-parameter dependent systems. In: IEEE Conf. on Decision and Control, New Orleans, Louisiana, pp. 2743–2748 (2007)

64. Fuller, A.T.: Conditions for a matrix to have only characteristic roots with negative real parts. Journal of Mathematical Analysis and Applications 23, 71–98 (1968)

65. Gahinet, P., Apkarian, P., Chilali, M.: Affine parameter-dependent Lyapunov functions and real parametric uncertainty. IEEE Trans. on Automatic Control 41(3), 436–442 (1996)

66. Garulli, A., Masi, A., Vicino, A.: Convex relaxations for quadratic distance problems. In: IEEE Conf. on Decision and Control, Cancun (Mexico), December 2008, pp. 5444–5449 (2008)

67. Genesio, R., Tartaglia, M., Vicino, A.: On the estimation of asymptotic stability regions: State of the art and new proposals. IEEE Trans. on Automatic Control 30, 747–755 (1985)

68. Geromel, J.C., Colaneri, P.: Robust stability of time varying polytopic systems. Systems and Control Letters 55(1), 81–85 (2006)

69. Goebel, R., Teel, A.R., Hu, T., Lin, Z.: Conjugate convex Lyapunov functions for dual linear differential inclusions. IEEE Transactions on Automatic Control 51(4), 661–666 (2006)

70. Hardy, G., Littlewood, J.E., Pólya, G.: Inequalities, 2nd edn. Cambridge University Press, Cambridge (1988)

71. Henrion, D., Garulli, A. (eds.): Positive Polynomials in Control. Lecture Notes in Control and Information Sciences, vol. 312. Springer, London (2005)

72. Henrion, D., Lasserre, J.-B.: GloptiPoly: Global optimization over polynomials with Matlab and SeDuMi. ACM Trans. on Mathematical Software 29(2), 165–194 (2003)

73. Henrion, D., Lasserre, J.-B.: Detecting global optimality and extracting solutions in GloptiPoly. In: Henrion, D., Garulli, A. (eds.) Positive Polynomials in Control. Lecture Notes in Control and Information Sciences, vol. 312. Springer, London (2005)

74. Henrion, D., Lasserre, J.-B.: Convergent relaxations of polynomial matrix inequalities and static output feedback. IEEE Trans. on Automatic Control 51(2), 192–202 (2006)

75. Henrion, D., Sebek, M., Kucera, V.: Positive polynomials and robust stabilization with fixed-order controllers. IEEE Trans. on Automatic Control 48, 1178–1186 (2003)

76. Hershkowitz, D.: Recent directions in matrix stability. Linear Alg. Appl. 171, 161–186 (1992)

77. Hol, C.W.J., Scherer, C.W.: Computing optimal fixed order $\mathscr{H}_\infty$-synthesis values by matrix sum of squares relaxations. In: IEEE Conf. on Decision and Control, Paradise Island, Bahamas (2004)

78. Hol, C.W.J., Scherer, C.W.: A sum-of-squares approach to fixed order H_∞-synthesis. In: Henrion, D., Garulli, A. (eds.) Positive polynomials in Control. Lecture Notes in Control and Information Sciences, pp. 45–71. Springer, Heidelberg (2005)

79. Horisberger, H.P., Belanger, P.R.: Regulators for linear time invariant plants with uncertain parameters. IEEE Trans. on Automatic Control 21(5), 705–708 (1976)

80. Hu, T., Lin, Z.: Composite quadratic Lyapunov functions for constrained control systems. IEEE Transactions on Automatic Control 48(3), 440–450 (2003)

81. Jarvis-Wloszek, Z., Feeley, R., Tan, W., Sun, K., Packard, A.: Control applications of sum of squares programming. In: Henrion, D., Garulli, A. (eds.) Positive polynomials in Control. Lecture Notes in Control and Information Sciences, pp. 3–22. Springer, Heidelberg (2005)

82. Jarvis-Wloszek, Z., Packard, A.K.: An LMI method to demonstrate simultaneous stability using non-quadratic polynomial Lyapunov functions. In: IEEE Conf. on Decision and Control, Las Vegas, Nevada, pp. 287–292 (2002)

83. Johansson, M., Rantzer, A.: Computation of piecewise quadratic Lyapunov functions for hybrid systems. IEEE Trans. on Automatic Control 43, 555–559 (1998)

84. Khalil, H.K.: Nonlinear Systems. McMillan Publishing Company, New York (1992)

85. Lasserre, J.-B.: Global optimization with polynomials and the problem of moments. SIAM Journal of Optimization 11(3), 796–817 (2001)

86. Lasserre, J.-B., Netzer, T.: Sos approximations of nonnegative polynomials via simple high degree perturbations. Math. Zeitschrift 256, 99–112 (2006)

87. Lasserre, J.-B., Netzer, T.: Sufficient conditions for a polynomial to be a sum of squares. Arch. Math. 89, 390–398 (2007)

88. Lavaei, J., Aghdam, A.G.: Robust stability of LTI systems over semi-algebraic sets using sum-of-squares matrix polynomials. IEEE Trans. on Automatic Control 53(1), 417–423 (2008)

89. Lavaei, J., Sojoudi, S., Aghdam, A.: Rational Optimization using Sum-of-Squares Techniques. In: IEEE Conf. on Decision and Control, New Orleans (December 2007)

90. Leite, V.J.S., Peres, P.L.D.: An improved LMI condition for robust $\mathscr{D}$-stability of uncertain polytopic systems. IEEE Trans. on Automatic Control 48(3), 500–504 (2003)

91. Lofberg, J.: YALMIP: A toolbox for modeling and optimization in MATLAB. In: CACSD Conf., Taipei, Taiwan (2004)

92. Margaliot, M., Langholz, G.: Necessary and sufficient conditions for absolute stability: the case of second-order systems. IEEE Trans. on Circuits and Systems I 50(2), 227–234 (2003)

93. Montagner, V.F., Peres, P.L.D.: A new LMI condition for the robust stability of linear time-varying systems. In: IEEE Conf. on Decision and Control, Maui, Hawaii, pp. 6133–6138 (2003)

94. Montagner, V.F., Peres, P.L.D.: Robust stability and $\mathscr{H}_\infty$ performance of linear time-varying systems in polytopic domains. International Journal of Control 77(15), 1343–1352 (2004)

95. Narendra, K.S., Taylor, J.H.: Frequency domain criteria for absolute stability. Academic Press Inc., New York (1973)
96. Narendra, K.S., Tripathi, S.S.: Identification and optimization of aircraft dynamics. Journal of Aircraft 10, 193–199 (1973)
97. Nesterov, Y.: Squared functional systems and optimization problems. In: Frenk, H., Roos, K., Terlaky, T., Shang, S. (eds.) High Performance Optimization. Kluwer, Norwell (2000)
98. Nesterov, Y., Nemirovsky, A.: Interior-Point Polynomial Methods in Convex Programming. SIAM, Philadelphia (1994)
99. Neto, A.T.: Parameter dependent Lyapunov functions for a class of uncertain linear systems: an LMI approach. In: IEEE Conf. on Decision and Control, Phoenix, Arizona, pp. 2341–2346 (1999)
100. Oishi, Y.: An asymptotically exact approach to robust semidefinite programming problems with function variables. In: Chesi, G., Henrion, D. (eds.) Special Issue on Positive Polynomials in Control. IEEE Trans. on Automatic Control, vol. 54, pp. 1000–1006 (2009)
101. Olas, A.: Construction of optimal Lyapunov function for systems with structured uncertainties. IEEE Trans. on Automatic Control 39, 167–171 (1994)
102. Oliveira, R.C.L.F., de Oliveira, M.C., Peres, P.L.D.: Convergent LMI relaxations for robust analysis of uncertain linear systems using lifted polynomial parameter-dependent Lyapunov functions. Systems and Control Letters 57(8), 680–689 (2008)
103. Oliveira, R.C.L.F., Peres, P.L.D.: Parameter-dependent LMIs in robust analysis: Characterization of homogeneous polynomially parameter-dependent solutions via LMI relaxations. IEEE Trans. on Automatic Control 52(7), 1334–1340 (2007)
104. Papachristodoulou, A.: Analysis of nonlinear time delay systems using the sum of squares decomposition. In: American Control Conf., Boston, MA (2004)
105. Papachristodoulou, A., Prajna, S.: A tutorial on sum of squares techniques for system analysis. In: American Control Conf., Portland, Oregon, pp. 2686–2700 (2005)
106. Parrilo, P.A.: Structured semidefinite programs and semialgebraic geometry methods in robustness and optimization. PhD thesis, California Institute of Technology (2000)
107. Parrilo, P.A.: Semidefinite programming relaxations for semialgebraic problems. Math. Programming Ser. B 96(2), 293–320 (2003)
108. Peaucelle, D., Arzelier, D., Bachelier, O., Bernussou, J.: A new robust $\mathscr{D}$-stability condition for real convex polytopic uncertainty. Systems and Control Letters 40, 21–30 (2000)
109. Peaucelle, D., Ebihara, Y., Arzelier, D., Hagiwara, T.: General polynomial parameter-dependent Lyapunov functions for polytopic uncertain systems. In: International Symposium on Mathematical Theory of Networks and Systems, Kyoto, Japan, pp. 2238–2242 (2006)
110. Peaucelle, D., Sato, M.: LMI tests for positive definite polynomials: Slack variable approach. IEEE Trans. on Automatic Control 54(4), 886–891 (2009)
111. Peet, M., Papachristodoulou, A., Lall, S.: Positive forms and stability of linear time-delay systems. In: IEEE Conf. on Decision and Control, San Diego, CA, pp. 187–192 (2006)
112. Peres, P.L.D., Souza, S.R., Geromel, J.C.: Optimal $\mathscr{H}_2$ control for uncertain systems. In: American Control Conf. (1992)
113. Powers, V., Wörmann, T.: An algorithm for sums of squares of real polynomials. Journal of Pure and Applied Linear Algebra 127, 99–104 (1998)
114. Prajna, S., Jadbabaie, A.: Safety verification of hybrid systems using barrier certificates. In: Alur, R., Pappas, G.J. (eds.) HSCC 2004. LNCS, vol. 2993, pp. 477–492. Springer, Heidelberg (2004)

115. Prajna, S., Papachristodoulou, A.: Analysis of switched and hybrid systems - beyond piecewise quadratic methods. In: American Control Conf., Denver, CO (2003)

116. Prajna, S., Papachristodoulou, A., Parrilo, P.A.: SOSTOOLS: a general purpose sum of squares programming solver. In: IEEE Conf. on Decision and Control, Las Vegas, Nevada (2002)

117. Prajna, S., Parrilo, P.A., Rantzer, A.: Nonlinear control synthesis by convex optimization. IEEE Trans. on Automatic Control 49(2), 310–314 (2004)

118. Putinar, M.: Positive polynomials on compact semi-algebraic sets. Ind. Univ. Math. 42, 969–984 (1993)

119. Radziszewski, B.: O najlepszej funkcji lapunowa. Technical report, IFTR Reports (1977)

120. Ramos, D.C.W., Peres, P.L.D.: A less conservative LMI condition for the robust stability of discrete-time uncertain systems. Systems and Control Letters 43, 371–378 (2001)

121. Reznick, B.: Extremal psd forms with few terms. Duke Mathematical Journal 45(2), 363–374 (1978)

122. Reznick, B.: Some concrete aspects of Hilbert's 17th problem. Contemporary Mathematics 253, 251–272 (2000)

123. Reznick, B.: On hilberts construction of positive polynomials (submitted for publication) (2007), http://www.math.uiuc.edu/~reznick

124. Sakuwa, R., Fujisaki, Y.: Robust stability analysis of single-parameter dependent descriptor systems. In: IEEE Conf. on Decision and Control and European Control Conf., Seville, Spain, pp. 2933–2938 (2005)

125. Sato, M., Peaucelle, D.: Robust stability/performance analysis for linear time-invariant polynomially parameter dependent systems using polynomially parameter-dependent Lyapunov functions. In: IEEE Conf. on Decision and Control, San Diego, CA, pp. 5807–5813 (2006)

126. Scherer, C.W.: Relaxations for robust linear matrix inequality problems with verifications for exactness. SIAM J. Matrix Anal. Appl. 27(1), 365–395 (2005)

127. Scherer, C.W.: LMI relaxations in robust control. European Journal of Control 12(1), 3–29 (2006)

128. Scherer, C.W., Hol, C.W.J.: Matrix sum-of-squares relaxations for robust semi-definite programs. Math. Programming Ser. B 107(1-2), 189–211 (2006)

129. Schmuedgen, K.: The K-moment problem for compact semi-algebraic sets. Mathematische Annalen 289, 203–206 (1991)

130. Seiler, P.: Stability region estimates for SDRE controlled systems using sum of squares optimization. In: American Control Conf., vol. 3, pp. 1867–1872 (2003)

131. Shor, N.Z.: Class of global minimum bounds of polynomial functions (in Russian). Cybernetics 11-12, 731–734 (1987)

132. Shor, N.Z.: Nondifferentiable optimization and polynomial problems. Kluwer, Dordrecht (1998)

133. Stengle, G.: A nullstellensatz and a positivstellensatz in semialgebraic geometry. Math. Ann. 207, 87–97 (1974)

134. Sturm, J.F.: Using SeDuMi 1.02, a MATLAB toolbox for optimization over symmetric cones. Optimization Methods and Software 11-12, 625–653 (1999)

135. Tan, W., Packard, A.: Stability region analysis using polynomial and composite polynomial Lyapunov functions and sum-of-squares programming. IEEE Trans. on Automatic Control 53(2), 565–571 (2008)

136. Tesi, A., Vicino, A.: Robust stability of state space models with structured uncertainties. IEEE Trans. on Automatic Control 35, 191–195 (1990)

137. Tibken, B.: Estimation of the domain of attraction for polynomial systems via LMI's. In: IEEE Conf. on Decision and Control, Sydney, Australia, pp. 3860–3864 (2000)

138. Topcu, U., Packard, A., Seiler, P.: Local stability analysis using simulations and sum-of-squares programming. Automatica 44(10), 2669–2675 (2008)

139. Šiljak, D.D.: Nonlinear Systems: Parametric Analysis and Design. John Wiley & Sons, New York (1969)

140. Šiljak, D.D.: Parameter space methods for robust control design: a guided tour. IEEE Trans. on Automatic Control 34(7), 674–688 (1989)

141. van der Schaft, A.J.: $\mathscr{L}_2$-gain analysis of nonlinear state feedback $\mathscr{H}_\infty$-control. IEEE Trans. on Automatic Control 37(6), 770–784 (1992)

142. Vandenberghe, L., Boyd, S.: Semidefinite programming. SIAM Review 38, 49–95 (1996)

143. Vicino, A., Tesi, A., Milanese, M.: Computation of nonconservative stability perturbation bounds for systems with nonlinearly correlated uncertainties. IEEE Trans. on Automatic Control 35, 835–841 (1990)

144. Wang, F., Balakrishnan, V.: Improved stability analysis and gain-scheduled controller synthesis for parameter-dependent systems. IEEE Trans. on Automatic Control 47(5), 720–734 (2002)

145. Wu, F., Prajna, S.: Sos-based solution approach to polynomial lpv system analysis and synthesis problems. Int. Journal of Control 78(8), 600–611 (2005)

146. Xie, L., Shishkin, S., Fu, M.: Piecewise Lyapunov functions for robust stability of linear time-varying systems. Systems and Control Letters 31, 165–171 (1997)

147. Xu, J., Xie, L.: Homogeneous polynomial Lyapunov functions for piecewise affine systems. In: American Control Conf., pp. 581–586 (2005)

148. Zelentsovsky, A.L.: Nonquadratic Lyapunov functions for robust stability analysis of linear uncertain systems. IEEE Trans. on Automatic Control 39(1), 135–138 (1994)

149. Zhang, X., Tsiotras, P., Iwasaki, T.: Parameter-dependent Lyapunov functions for exact stability analysis of single parameter dependent LTI systems. In: IEEE Conf. on Decision and Control, Maui, Hawaii, pp. 5168–5173 (2003)

150. Zhou, K., Khargonekar, P.P.: Stability robustness bounds for linear state-space models with structured uncertainties. IEEE Trans. on Automatic Control 32, 621–623 (1987)

Authors' Biography

Graziano Chesi

Graziano Chesi received the Laurea in Information Engineering from the Università di Firenze, Italy, in 1997, and the Ph.D. in Systems Engineering from the Università di Bologna, Italy, in 2001. From 2000 to 2006 he was with the Dipartimento di Ingegneria dell'Informazione of the Università di Siena, Italy, and in 2006 he joined the Department of Electrical and Electronic Engineering of the University of Hong Kong, Hong Kong Special Administrative Region. He was a visiting scientist at the Department of Engineering of the University of Cambridge, UK, from 1999 to 2000, and at the Department of Information Physics and Computing of the University of Tokyo, Japan, from 2001 to 2004.

Dr. Chesi was the recipient of the Best Student Award of the Facoltà di Ingegneria of the Università di Firenze in 1997. He is in the Editorial Board of the IEEE Conference on Decision and Control and other international conferences, moreover he is Associate Editor of *Automatica*, Associate Editor of the *IEEE Transactions on Automatic Control*, and Guest Editor of the Special Issue on *Positive Polynomials in Control* of the *IEEE Transactions on Automatic Control*. He is the first author of more than 90 technical publications. He is the Founder and Chair of the Technical Committee on Systems with Uncertainty of the IEEE Control Systems Society. His research interests include computer vision, nonlinear systems, optimization, robotics, robust control, and systems biology.

Dr. Graziano Chesi
Department of Electrical and Electronic Engineering
University of Hong Kong
Pokfulam Road, Hong Kong
Phone: +852-22194362
Fax: +852-25598738
Email: chesi@eee.hku.hk
Homepage: http://www.eee.hku.hk/~chesi

Andrea Garulli

Andrea Garulli was born in Bologna, Italy, in 1968. He received the Laurea in Electronic Engineering from the Università di Firenze, Italy, in 1993, and the Ph.D. in System Engineering from the Università di Bologna in 1997. In 1996 he joined the Dipartimento di Ingegneria dell'Informazione of the Università di Siena, Italy, where he is currently Professor of Control Systems.

He has been member of the Conference Editorial Board of the IEEE Control Systems Society and Associate Editor of the *IEEE Transactions on Automatic Control*. He currently serves as Associate Editor for the *Journal of Control Science and Engineering*. He is author of more than 120 technical publications, Editor of the books *Robustness in Identification and Control* (Springer, 1999) and *Positive Polynomials in Control* (Springer, 2005), and Guest Editor of the Special Issue *Robustness in Identification and Control* of the *International Journal of Robust and Nonlinear Control* (2001). His present research interests include system identification, robust estimation and filtering, robust control, mobile robotics and autonomous navigation.

Prof. Andrea Garulli
Dipartimento di Ingegneria dell'Informazione
Università di Siena
Via Roma, 56
53100 Siena, Italy
Phone: +39-0577-233612
Fax: +39-0577-233602
Email: garulli@dii.unisi.it
Homepage: http://www.dii.unisi.it/~garulli

Alberto Tesi

Alberto Tesi was born in 1957, received the Laurea degree in Electronics Engineering from the Università di Firenze, in 1984, and the Ph.D. degree in Systems Engineering from the Università di Bologna, in 1989. In 1990 he joined the Dipartimento di Sistemi e Informatica of the Università di Firenze, where he is currently a Professor of Automatic Control. Since 2006 he is the Dean of the Faculty of Engineering. He has been elected Rector of the Università di Firenze for the period 2009-2013.

He was Associate Editor of the *IEEE Transactions on Circuits and Systems* (1994-1995) and of the *IEEE Transactions on Automatic Control* (1995-1998). Actually, he is Associate Editor of *Systems and Control Letters*. He was member of the Conference Editorial Board of the Conference on Decision and Control (1994-1999) and the American Control Conference (1995-2000), and member of the program committee of several international conferences. Since 2008 he is member of the Policy Committee of the International Federation of Automatic Control (IFAC). His research interests are mainly in analysis of nonlinear dynamics of complex systems, robust control of linear systems and optimization. He is author of about 170 scientific publications.

Prof. Alberto Tesi
Dipartimento di Sistemi e Informatica
Facoltà di Ingegneria
Università di Firenze
Via di Santa Marta, 3
50139 Firenze, Italy
Phone: +39-055-4796263
Fax: +39-055-4796363
Email: alberto.tesi@unifi.it

Antonio Vicino

Antonio Vicino was born in 1954. He received the Laurea in Electrical Engineering from the Politecnico di Torino, Torino, Italy, in 1978. From 1979 to 1982 he held several Fellowships at the Dipartimento di Automatica e Informatica of the Politecnico di Torino. He was assistant professor of Automatic Control from 1983 to 1987 at the same Department. From 1987 to 1990 he was Associate Professor of Control Systems at the Università di Firenze. In 1990 he joined the Dipartimento di Ingegneria Elettrica, Università di L'Aquila, as Professor of Control Systems. Since 1993 he is with the Università di Siena, where he founded the Dipartimento di Ingegneria dell'Informazione and covered the position of Head of the Department from 1996 to 1999. From 1999 to 2005 he was Dean of the Engineering Faculty. In 2000 he founded the Centro per lo Studio dei Sistemi Complessi (CSC) of the University of Siena, where he presently covers the position of Director.

He has served as Associate Editor for the *IEEE Transactions on Automatic Control* from 1992 to 1996 and for the *IEEE Transactions on Circuits and Systems II* from 2006 to 2007. Presently he serves as Associate Editor for *Automatica*, and Associate Editor at Large for *IEEE Transactions on Automatic Control*. He is Fellow of the IEEE and of the IFAC. He is author of more than 230 technical publications, Editor of 2 books on *Robustness in Identification and Control*, Guest Editor of the Special Issue *Robustness in Identification and Control* of the *International Journal of Robust and Nonlinear Control* and of the Special Issue *System Identification* of the *IEEE Transactions on Automatic Control*. He has worked on stability analysis of nonlinear systems and time series analysis and prediction. Presently, his main research interests include robust control of uncertain systems, robust identification and filtering, mobile robotics and applied system modeling.

Prof. Antonio Vicino
Dipartimento di Ingegneria dell'Informazione
Università di Siena

Via Roma, 56
53100 Siena, Italy
Phone: +39-0577-233615
Fax: +39-0577-233602
Email: vicino@dii.unisi.it
Homepage: http://www.dii.unisi.it/~vicino

Index

Lecture Notes in Control and Information Sciences

Edited by M. Thoma, F. Allgöwer, M. Morari

Further volumes of this series can be found on our homepage:
springer.com